Praise for
A Bigger Picture

"In this moment of intersecting crises, Vanessa Nakate continues to teach a most critical lesson. She reminds us that while we may all be in the same storm, we are not all in the same boat."

—Greta Thunberg

"Through Vanessa Nakate's eyes, *A Bigger Picture* shows us the threat of climate change to people in East Africa and the relentless courage of one activist fighting to be heard. Vanessa is more than an inspiration—she's an indispensable voice for our future."

—Malala Yousafzai

"Vanessa Nakate is a powerful global voice. A strong spirit who will clearly not give up and only grow in strength."

—Angelina Jolie

"Vanessa Nakate's message couldn't be more urgent or her voice more desperately needed. At once intimate and sweeping, *A Bigger Picture* is a must-read for anyone who cares about the future."

—Elizabeth Kolbert, author of *Under a White Sky* and *The Sixth Extinction*

"This is a wonderful story, wonderfully told! Vanessa Nakate is a crucial climate leader, reminding us of one of the iron laws of global warming: the less you did to cause it, the sooner and harder you get hit. Thank heaven her voice will echo far and wide, and down through the years."

—Bill McKibben, author of *Falter: Has the Human Game Begun to Play Itself Out?*

"Vanessa's story, voice, and fearless spirit are an inspiration to all of us. This book is a vital reminder that the costs of climate change have been transforming negatively the lives of those who have had the least part in causing the problem. Without racial justice and equality, climate justice can never be a reality."

—Mary Robinson, former president of Ireland and author of *Climate Justice: Hope, Resilience, and the Fight for a Sustainable Future*

A Bigger Picture

MY FIGHT TO BRING A NEW AFRICAN VOICE TO THE CLIMATE CRISIS

VANESSA NAKATE

MARINER BOOKS

New York   Boston

A BIGGER PICTURE. Copyright © 2021 by Vanessa Nakate. All
rights reserved. Printed in the United States of America. No part
of this book may be used or reproduced in any manner whatsoever
without written permission except in the case of brief quotations
embodied in critical articles and reviews. For information, address
HarperCollins Publishers, 195 Broadway, New York, NY 10007.

HarperCollins books may be purchased for educational, business,
or sales promotional use. For information, please email the
Special Markets Department at SPsales@harpercollins.com.

First published in 2021 by One Boat, an imprint of Pan Macmillan.

A hardcover edition of this book was published in
2021 by Houghton Mifflin Harcourt.

FIRST MARINER BOOKS PAPERBACK EDITION PUBLISHED 2022.

Library of Congress Cataloging-in-Publication Data has been applied for.

ISBN 978-0-06-326912-5 (pbk.)

22 23 24 25 26   LSCC   10 9 8 7 6 5 4 3 2 1

Contents

To the People and the Planet

Introduction

I couldn't believe what I was seeing—or rather, what I wasn't.

It was a freezing cold day in January 2020, and I was scrolling through my social media feeds. I'd just finished lunch with other climate activists, who like me were in Davos, Switzerland, to urge some of the three thousand business leaders, financiers, politicians, opinion formers, celebrities, and other globetrotters attending the annual World Economic Forum (WEF) to get serious about the climate crisis. We'd held a press conference that morning, before which I'd posed for cameras with four other activists, and I'd stepped away from the dining area to find out how the media was reporting our message.

Within a minute, I came upon a link to an article that featured one of the photos that had been taken of us. My heart nearly stopped. It was clearly the picture I'd been in, since you could make out the edge of my coat on the far left of the frame. But I was nowhere to be seen. I'd been cropped out.

I cycled rapidly through my feelings. I was frustrated, angry, and embarrassed. As I looked at the image, it became impossible to ignore that of the five women who'd posed for that photo, I was the only one who wasn't from Europe and the only one who was Black. They hadn't just cropped me out, I realized. They'd cropped out a whole continent.

At the press conference that morning in Davos, I'd been the only climate activist from Africa (there were a few others at the WEF itself), and not only had I been cut out of the Associated Press's photo but out of the AP's article that reported on our press conference too. "Does that mean I have no value as an activist or the people from Africa don't have any value at all?" I asked in a ten-minute video I streamed live later that day. I was struck by the cruel irony of the exclusion of the only African from the photo. "We don't deserve this," I said. "Africa is the least emitter of carbons, but we are the most affected by the climate crisis."

For a year, I'd organized climate strikes on the streets of Kampala, the capital and largest city in Uganda, in east-central Africa, where I live, to demand action on the climate emergency. I'd attended international climate conferences and been active online, and now I'd come to Davos to help more people wake up to the truth that global heating is not an abstraction or a theoretical event awaiting the planet in a few decades.

My message was, and is, straightforward: People in Uganda, in Africa, and across what's called the Global South are losing their homes, their harvests, their incomes, even their lives, and any hopes of a livable future *right now*.

This situation is not only terrible, it's also unjust. Although the African continent has just 15 percent of the world's population, it is responsible for only between 2 and 3 percent of global energy-related carbon dioxide emissions.[1] The average African's greenhouse gas emissions are a fraction of those of people living in the US, Europe, China, the United Arab Emirates, Australia, or many other countries. An Oxfam study concluded that a person in the UK will have emitted more CO_2 in the first two weeks of 2020 than someone in Uganda or six other African countries will in the whole year.[2]

Nonetheless, Africa will, according to the African Development Bank, bear almost half the costs of adapting to the consequences of climate change, and seven of the ten countries most susceptible to the harshest effects of the climate crisis are in Africa: South Sudan, Nigeria, Ethiopia, Eritrea, Chad, Sierra Leone, and the Central African Republic.

Those with the fewest resources and who've contributed the least to the crisis are contending with the gravest consequences: more frequent and more serious flooding, longer droughts, periods of extreme heat, and rising sea levels. Increased food scarcity, forced migration, economic losses, and higher rates of death are also disproportionately affecting people of color, not only across Africa and the rest of the Global South, but in the Global North too.[3]

This is my world—a world where Earth's temperature has already risen 1.2°C (2.16°F) above pre-industrial levels. A planet that's 2°C hotter is a death sentence for countries like Uganda. Yet, as you read this, we're on course for temperature rises that are much, much more than 2°C. That means many more millions of people will be displaced and extreme weather events will strain health and economic systems to the breaking point. At the same time, the world's oceans are being depleted, biodiversity is collapsing, and species are going extinct at a rate greater than since the time of the dinosaurs.

My video response was seen by tens of thousands of people around the world, including many in Uganda, who shared my outrage and disappointment. Like me, they realized that, quite literally, something was very wrong with this picture. Being cropped out of that photo changed the course of my activism and my life. It reframed my thoughts about race, gender, equity, and climate justice; and it led to the words you're now reading.

In *A Bigger Picture*, I explain why that photo and that moment mattered, and why it's crucial that the fight against climate change includes voices like mine. I describe how I first became a climate striker, and my eventual journey to the Alps and what has happened since. I show how what we must call the climate emergency is an immediate, even daily struggle for millions of people, including across Africa, and how the heating of Earth's atmosphere is connected to everything: economics, society, politics, and many forms of inequality and injustice—racial, gender, and geographic.

Like many of the young climate activists I've organized with and been inspired by, I live in a profoundly interconnected world, with instant access to huge amounts of information (and disinformation) and more means of connecting to others than at any time in history. Those of us born at the end of the last century and in the early years of this one have grown up in the shadow of HIV/AIDS, terrorism, financial meltdowns, and huge technological change and disruption. We've witnessed greater concentrations of wealth and increased disparities of power. Many of us have experienced firsthand how our planet's ecosystems are breaking down under climatic stresses unprecedented in human history.

Perhaps more than any other age group, we are questioning the premise of an economic, social, and political model that has led us to a precipice beyond which *no* economic or governance system will survive. These realities have shaped our recognition that we, and those that follow us, will bear the brunt of several centuries of burning fossil fuels and our calamitous failure to leave the remaining carbon in the ground.

A Bigger Picture also showcases the work and perspectives of a fresh wave of activists from a new generation. Many of them focus their vision on and from Africa, a continent that has been ignored, silenced, and exploited for too long. We believe that at the center of this effort must be a genuine commitment not only to environmental, racial, and climate justice, but to the empowerment of girls and women, who are facing the crisis most acutely and are at the forefront of efforts to combat it. Without tackling climate change, we won't be able to achieve the United Nations' Sustainable Development Goals, or bring about a resilient and sustainable future. I also share the practical solutions that climate activists are applying to support communities in Uganda and other countries in Africa and around the world.

Finally, I offer ideas for how you can become active in addressing the climate emergency wherever you live, and how you can amplify the voices and acknowledge the presence of those who've too often been left out of the picture.

*   *   *

I wrote *A Bigger Picture* in the midst of the COVID-19 pandemic and, like you, I am stunned and deeply saddened at the loss of so many people in so many countries to the virus. Across the world, families, communities, and nations are in shock and are mourning the livelihoods ruined, the families dislocated, the schooling interrupted or curtailed, and the businesses shuttered. We're also shaken by other shameful effects of the pandemic: the lack of access to health care and vaccines for people of color; the upturn in the incidence of child marriage and domestic violence; and the delaying of urgent action on the climate emergency. Though these inequities existed before COVID-19, the virus has brought them to the fore and made many of them worse.

In these multiple tragedies, we can find stark warnings and lessons. First, scientists are telling us that zoonotic diseases like COVID-19 will become more common in the future as we encroach on habitats where wild animals live; continue to use, raise, and sell wildlife in close proximity to human communities; and confine billions of domesticated animals in factory farms. Climate change is likely to increase the frequency and deadliness of such diseases.

Second, throughout the pandemic, people around the world have paid special care to the elderly, who've proved more vulnerable to the virus. We've kept them safe by staying inside. But for decades, many people in these generations have made decisions that will leave their heirs vulnerable to the effects of global heating.

Third, the pandemic has disproportionately affected those with fewer resources; less access to health care and enough nutritious food; more cramped living conditions; work that makes social distancing difficult; and underlying health conditions that put them at greater risk from the virus. A majority of these are people of color. This, too, echoes the climate crisis.

Finally, while governments have been telling us to follow the science on the coronavirus, they have not been following the science on climate change. They aren't moving nearly as fast or as comprehensively as scientists tell us we must to meet—or exceed—the commitments made under the 2015 Paris climate accord. The pandemic has reminded us that climate change is not in lockdown. It has demonstrated that we live in a deeply connected world and that we need one another to survive.

Even though the climate forecasts are terrifying, I still believe we can have hope. We have to. There isn't any other option. The pandemic has shown that (some) leaders *can* listen to the science, and the international community *can* act together with a common purpose. And, no matter how disturbing the present and future may appear, we have neither the time nor the luxury to shut down emotionally, especially those of us who live in countries where the climate crisis is a daily reality.

The stakes could not be higher: unless we take dramatic action now, whatever plans any of us have for the future—whether big or small—will fail. So, join me and some of the many young climate activists in Africa and around the world who are working *right now* to change that future. Let's fight together for what is right and what is just.

1

Finding My Cause

It's a long way both metaphorically and literally to Switzerland from Kampala, Uganda. If you'd told me in the summer of 2018 that I'd be a climate activist and in Davos eighteen months later, I wouldn't have known what you were talking about. *Where is Davos?* I would probably have asked, *And what is a climate activist?* So, the first thing you need to understand about me is that I'm as amazed by this journey as you may be.

I was twenty-two years old and approaching the end of my degree in Business Administration at Makerere University Business School (MUBS) in Kampala. MUBS was founded in 1997, and is a branch of Makerere University, the oldest, largest, and most well-regarded university in Uganda. I was starting to think about what to do once I graduated. The logical path would be for me to take a professional course at the Chartered Institute of Marketing, followed by a Master's in Business Administration or even a Doctorate in Marketing. Each credential would give me an advantage in the job market, which is very competitive.

In Uganda, there's a several-month gap between finishing your university course and the graduation ceremony. I had in mind contributing to society during that time by volunteering to help people in some way, but I wasn't sure how or what I wanted to do.

As it turned out, the answer was right in front of me.

During the spring, summer, and autumn of 2018, local news and my social media feeds filled with stories about massive floods devastating whole swathes of East Africa—from Djibouti and Somalia to Burundi and Rwanda. It was heartbreaking to see images of houses washed away, read about hundreds of people dying, and learn of many more displaced and in need of shelter, food, and medicine. Thousands of hectares of crops were being destroyed. In Kenya, which borders Uganda to the east, thousands of goats, sheep, and cows were killed. I saw images of little children wading through water that had turned red-brown as it filled with topsoil from the surrounding hillsides. The United Nations described the flooding in Somalia, where half a million people were affected, as the worst the region had ever seen.[1]

My own country wasn't spared. In May, Kalerwe and Bwaise, two informal settlements, flooded in Kampala, which is situated on the shores of Lake Victoria, Africa's largest lake, about 46 miles/70 km north of the Equator. In October, three days of relentless, heavy rain caused massive landslides in the mountainous regions of Bukalasi and Buwali in Bududa District in the east of the country. Fifty-one people died and twelve thousand were displaced. Many roads and four bridges were swept away. Tragically, in Maludu village, a landslide buried a primary (elementary) school in mud, and many children lost their lives.[2]

Meanwhile, in the arid Karamoja region of the far northeast, on the border of northern Kenya and South Sudan, the rains failed for a second year running. All this led Uganda's Ministry of Finance, Planning, and Economic Development to observe that

2018's droughts, irregular rainfall, and calamitous floods had "significantly impacted agriculture, hydro-electricity production, water resources, human settlements, and infrastructure." There would, the ministry added, be "long-term implications of persistent poverty and increased food insecurity."[3]

Uganda has a mainly warm, tropical climate apart from the mountainous regions, which can be much cooler. Its two rainy seasons run from March to May and September to November. In addition to Lake Victoria, from which the Nile River flows north, Uganda is fortunate to have many bodies of water, such as Lake Kyoga, and Lakes Albert and Edward, which we share with the Democratic Republic of Congo. It has ten national parks and ten percent of its land is forested, although that percentage is declining.

I knew that some parts of my country were prone to flooding, and that decades of deforestation had made landslides more likely. But something was different about the extreme events marking 2018. They seemed to be happening more frequently, occurring all over the country, lasting longer, and displaying greater ferocity. The rainy and dry seasons also appeared to have shifted and become more intense, with heavier rains, longer droughts, and sudden swings between the two.

I'd been taught about global warming in a module in geography at secondary (high) school. But in that class, the only one to even address the subject, the teacher had suggested that climate change was a problem we'd have to deal with in the future, and that it affected other parts of the world and other people. Could it be, I asked myself, that climate change wasn't in the future and elsewhere, but here and now: in Africa, in Uganda, in Kampala? Were these events I was learning about from the news while living at my parents' home in the capital city—inundations, record temperatures, failing harvests, hungry children, disease outbreaks, desperate refugees—going to be regular occurrences, the new normal? And what would happen if they grew worse? How many more harvests would be lost? How many more people would be displaced? And how many more would die?

At that time, I knew next to nothing about the world's response to climate change. I had no idea that in Paris in 2015, 197 countries had set a goal to reduce greenhouse gas emissions by 2100 so that the overall warming of the planet's atmosphere would be kept "well below" 2°C (3.6°F) above the levels it had been before the Industrial Revolution. In Paris, countries had also agreed to try to meet a more ambitious target to avoid the worst of the disruption researchers forecast: an increase in the global temperature of no more than 1.5°C (2.7°F).

But, I was to learn, in spite of the 2015 commitments, emissions hadn't declined and the Earth's temperature has already risen to 1.2°C (2.16°F) above pre-industrial levels. In fact, I found out, those commitments were nowhere near as sweeping as research by the United Nations' Intergovernmental Panel on Climate Change (IPCC) had demonstrated were essential. I read that not only were scientists telling us we had only a decade to decarbonize our economies before a temperature rise of 1.5°C or much higher was locked in, but the World Meteorological Association calculated there was a 20 percent chance that the global temperature would increase by 1.5°C as soon as 2024.[4] What was even more shocking was that the planet was on course for potentially a 3°C (5.4°F) temperature increase by 2050 and 7°C (12.6°F) by 2100—a civilization-ending scenario.[5]

I was stunned. Worry. Sadness. Fear. Anger. Bewilderment. Frustration. Disgust. These are some of the emotions expressed by scientists about the climate crisis on the Is This How You Feel? website.[6] As I watched the videos, listened to the podcasts, and read the blogs, social media posts, and newspaper articles, these emotions and more arose in me too.

And I had so many questions! Why wasn't climate change being more widely taught in our schools and universities? Why weren't we listening to the scientists? Why wasn't our Government acting? Why wasn't the international community working together more? What were all our leaders *doing*? Were we fooling ourselves in not taking this issue seriously?

December and January are some of the hottest months of the year in Uganda, and the Christmas and New Year period of 2018 was especially so. Some nights it remained so hot even after the sun went down that it was hard to sleep in my room in the attic. I decided to ask my Uncle Charles, who follows current events and reads a lot, whether he remembered it ever being quite as warm. "No," he replied. "Two or three decades ago, January was relatively temperate and wet." That meant it was perfect for harvesting maize (corn), cassava, beans, and sweet potatoes, all of which are grown during the rains of the previous months. "It's climate change, and yet no one is talking about it," he added.

Uncle Charles shook his head. "The farmers in our country have all been affected," he said. They may never have heard of the concept of global warming, he told me, but they could sense that something was wrong: "Farmers have seen the weather change, and they have to deal with the consequences." Then he said something that really captured my attention: "We need to do something about it, for the sake of the environment and for young people."

As I listened to Uncle Charles, my alarm and anger intensified. I began to see how accurate the term *climate crisis* actually was. In my research online, I discovered one person who recognized that it was just that: Swedish teenager Greta Thunberg, who'd started the Fridays For Future (FFF) movement. A few months before, Greta had begun to skip school each Friday and had instead stood outside the Swedish Parliament holding a sign, SKOLSTREJK FÖR KLIMATET ("School strike for climate"), to protest her country's (and the world's) failure to adequately address the warming of Earth's atmosphere.

I was impressed that here was someone much younger than I was who was bearing witness and taking to the streets. The more I researched, the more I discovered that other young people around the world, some not yet teenagers, were joining Greta in school strikes for the climate. With my Uncle Charles' words echoing in my head, I felt the pull to strike too. I began to feel like I *had* to become a climate activist. But I wasn't clear on what I'd actually do or how.

Several obstacles prevented me from acting. First, even though most people think of me as a friendly person, I'm actually quite shy and happy to spend time alone. I also don't like to draw attention to myself. How could I step out in this way as an "activist"? Who was I going to strike with? Should I hold a placard and stand at the corner nearest to my home, or was I going to be like Greta and strike in front of a government office or even the Ugandan Parliament, which I'd never even visited? Where to strike, it seemed, would be the hardest decision for me.

Another barrier I faced might be hard for some of you to understand. My society in Uganda, and others like it, has very strict set rules about what's appropriate for a young woman—rules that don't apply to men and boys. At the girls' boarding school I attended, we were taught to be demure and respectful of authority. A young woman standing by herself on the street with a sign is unusual and she may well be harassed, verbally bullied, or worse. She'd also likely be accused of being "desperate for a man"—an insult often hurled at single young women doing something unexpected—or even suspected of being a prostitute. If our parents or siblings heard we were protesting on the street, there's a good chance they'd be embarrassed or angry: "What were you thinking?!" they'd ask accusingly.

Yet another problem I faced was that it's hard to hold a public demonstration in Uganda. Permits for marches can be difficult to obtain administratively and the police can break up a large gathering of people and you can be arrested. I've seen it myself. Despite what the statutes may say, from what I've seen we don't really have free speech, and the authorities routinely ban rallies or protests that they deem "too political."

My contemporaries at Makerere University's main campus, across town from where I was studying at MUBS, frequently went on strike, especially over rises in tuition fees. Many of these protests had been broken up by the police using their batons and tear gas, which spread throughout the campus and even penetrated student housing. Students had been beaten and arrested. Sometimes the police even fired live bullets to try to disperse the crowd.

When the police in Uganda arrest you, they often take away all your possessions, including your phone. What if my climate strike led to my arrest? If the police confiscated my phone, who would even discover I'd been detained? Who would I tell? Who would bail me out of jail? How would my family react?

These thoughts and fears preoccupied me as the new year of 2019 began. Outwardly, you wouldn't have thought anything had changed. I kept working, as I had since I was twelve years old when I'd help out during the school holidays, in my family's battery supply shop. In most ways I was what you could call an ordinary college student. I occasionally went to parties, where I loved to dance and talk with friends. In the evenings, I enjoyed *telenovelas* from Latin America dubbed into English on Telemundo, watched Ugandan talent shows and beauty pageants, and often listened to the music of One Direction, Ed Sheeran, Taylor Swift, and Ugandan rapper Fik Fameica. I was also looking forward to my graduation, which was coming up later in January. It had been an intense three years at university, because I was studying full-time and was now part-time manager at the shop, handling finances and writing receipts.

Even though I'm the eldest of five children and close to my mom and dad, I've never been the kind of person who confides in her parents. While at MUBS, I lived in my parents' house in the southeastern part of Kampala. Yes, I would have liked to have had the experience of living in a university hostel, as several of my friends did, but since MUBS was close to home, my parents decided it wasn't worth the expense. Although I didn't have as much privacy as some university students who live in a hostel, because my room was in the attic, I could plan the strike without anyone knowing what I had in mind.

I mainly kept my thoughts about striking to myself because I didn't want my parents to try to dissuade me from holding a public climate strike. It's not that they wouldn't have understood my wish to speak out. As we were growing up, my parents had encouraged me and my two sisters and two brothers to do what we thought was right, and not just what was popular. They had always made clear that if my siblings and I wanted to undertake a project, and it was for a positive cause and made us happy, then we should do it to the best of our abilities and see where it would lead. It was more the reaction from my female classmates that I worried about: "You're going against every-thing we were taught!" I imagined my school friends saying, if they found out I'd be standing on the street with a sign.

And yet . . . I couldn't stop replaying in my mind the differ-ent student strikers I saw on social media. Many of them were young women or girls. Alexandria Villaseñor, an American, was only fourteen years old and had founded Earth Uprising, a youth climate network. Lilly Platt had started a campaign to encourage people to clean up plastic waste in the Netherlands in 2015, when she was only seven. And, of course, there was Greta. If they had the confidence to go out in public and call for climate action, I told myself, surely I, a soon-to-be university graduate in a country where the consequences of the climate crisis were right before my eyes, could join them. If I didn't, would I ever forgive myself?

By early January, I'd spent many hours weighing up whether I should begin climate activism or not. I'd wasted precious time, scared about what might happen and worried about what other people—my friends, my family, *anyone*—would think or say about me. I knew I'd delayed too long already. It was time for me to leave that place of fear and face the world.

* * *

On Saturday, January 5, 2019, I told my brothers, who were home from boarding school, Paul Christian (then aged 14) and Trevor (aged 10), my two cousins who were visiting, Nathan (aged 11) and Varak (aged 9), and a third cousin, Isabella, who is my age and lives with my family in Kampala, that I wanted us to strike.[7]

"What are we going to strike for?" Isabella asked.

"For the climate," I replied.

Nathan asked why, and I tried to explain, for their sakes and my own. "We're going to demand climate action," I told them. "We want the politicians and business community to do something." I wasn't sure I was convincing them. I added, "We're doing it on behalf of the people who are suffering because of climate disasters." Then I looked at my four young relatives, their faces expectant and Isabella's skeptical. "And I want to have my very first strike with you."

I could see they were getting excited. As the oldest child in my family, I was used to giving my siblings some direction. This came in handy now. "So, we have to make our placards," I told the four young ones. What kids wouldn't enjoy that? Isabella and I could join in and supervise.

Earlier that day, I'd bought markers so we could write in big, bold letters. Luckily, one of my sisters is a very good artist, and she owned a large book of art paper, which she said we could use. We settled down to make the signs we'd carry.

"What shall we write?" Varak, the nine-year-old, asked.

I wanted us to express something positive, and to ensure that my younger family members held placards they themselves would understand. We decided to pick slogans we thought wouldn't be too threatening, and so wrote several, in English. TREES ARE IMPORTANT FOR US; NATURE IS LIFE; WHEN YOU PLANT A TREE, YOU PLANT A FOREST; THANKS FOR THE GLOBAL WARMING (that was our sarcastic one); and CLIMATE STRIKE NOW. We also drew some trees next to the letters.

As we scribbled away in the living room, my mom grew curious and poked her head around the doorway. "Mwe mukolakyi? Kulwakyi muwandika ebigambo ku mpapula?" she asked in Luganda, my mother tongue: *What are you doing? Why are you writing on all these pieces of paper?*

I decided to be straightforward. "We're going on a climate strike," I replied.

"What is *that*?" she asked. This was completely new to her, as it was to all of us.

"We're fighting for the protection of the environment. We need these signs," I said. "We want to make sure the Government does something about climate change."

Mom paused as she considered what I'd told her. "That's a good thing," she said. "But aren't you worried, going to the streets?"

"I've been scared this whole time," I responded. "But I've decided we need to do this."

I could see she was still anxious. "Since this is a peaceful strike, and I'm with the kids," I added, "I don't think you need to worry."

My mother nodded. I wasn't sure she was convinced, but she just said, "OK."

The next morning, the six of us got up early and left home at about seven o'clock, slipping out of the door so as not to wake up my parents. The day was sunny but with a cold bite in the air, so we all wore sweaters.

Kampala, like many cities in Africa, has parts that are very modern, with paved roads, sidewalks, and tall buildings; and some areas that are green with trees and where the red soil is visible. Many areas are packed with traffic, with lots of noise from honking horns. I'd picked four places where we'd stand with our signs for thirty minutes each (I'd even set my phone alarm so we'd know when to leave). I'd chosen them strategically: they were in crowded markets at intersections with lots of traffic, where our messages could be seen by as many people as possible. Here, taxicabs, *matatus* (public minibuses seating up to fourteen passengers), and *boda bodas* (motorcycle taxis) compete for road space with cyclists and pedestrians. Each of the sites, in the neighborhoods of Kitintale and Bugolobi, wasn't far from our house, and they formed a kind of circle, allowing us to be close to home after we reached the last location.

As we walked to our first destination at Kitintale Market, we held our placards in the air, like a mini climate march. My brother Paul Christian took pictures of us so I could post images from the strike later on social media. Many people stared as we made our way along the sidewalk, clearly wondering what we were doing. At one point, a woman stopped in front of us and told us we should go to a site nearby where they were cutting down trees to make way for a school. "They need to understand they cannot do this," she said. "You can still have a school and let the trees remain standing." I said that I agreed, and made a mental note to follow up and find out more. (When I later visited the school, the head teacher wasn't there. I wrote a letter, asking to speak with the students and teachers about climate change, but never received a response. I did notice that the trees had been cut near the school, which was on my route to and from MUBS.)

On Sundays in Uganda, stalls in open-air markets sell fruits, vegetables, grains, and meat to shoppers, as well as cooked food for these customers, the stallholders themselves, and hungry passersby. Kitintale was no exception. More than a hundred vendors were setting up while customers were forming lines for the cooked food. As we stood holding our placards, my heart started to beat faster. How might people react?

Mostly, they just carried on with their Sunday morning

routines. Merchants arranged their bananas and peppers; a few shoppers looked in our direction and paused to read our signs. Nobody said anything directly to us, and no one yelled or cursed or tried to chase us away either. Still, I was so anxious I couldn't even feel my legs. Honestly, I wanted to leave as soon as possible. My cousin Varak told me later that she was also mortified, at least at the start. "Tubadde awo naye mbade njagala etakka limile!" she said. *I wanted the ground to swallow me up!*

Because it was our first one, that stop was the scariest and most stressful. But we stuck it out. I reminded myself that our messages had to be brought to the public regardless of how uncomfortable it was. In fact, when the alarm on my phone rang, announcing it was time to head to our next strike location, I was less worried about continuing the strike than I thought I'd be.

Bugolobi Stage was about a four-minute ride by *matatu* along busy Port Bell Road. It's called a "stage" because it's a depot and transit hub, so like a "staging" point for journeys. It's always bustling, with little stores selling refreshments and electronics, Internet access and money transfer services, phone-charging and clothes. A lot of people were here too. And while we received some curious glances, again nobody said anything. I could feel the strike getting easier. After another half an hour, we walked to the Village Mall, our third strike spot.

The Village Mall, also in Bugolobi, is an upscale, enclosed shopping area, where prices for everything from a cup of coffee to a piece of clothing are three times more expensive than they are elsewhere. By standing here we'd be seen not only by wealthier Ugandans, but also by white expatriates out for a meal or shopping. That was important. These people needed to see that climate activists are not only in Europe or the United States, but in Uganda as well. I'd also chosen the Mall because of the Shell

21

petrol (gas) station next door. I'd seen other climate activists targeting the oil industry. This location would allow us to draw drivers' attention to their complicity. We stood in front of the entrance and waved our placards, so that people in private vehicles, walkers, cyclists, bus-riders, and *matatu* passengers would see us as they passed by.

Our final destination was in Nakawa, in front of Capital Shoppers, a large supermarket across the road from my soon-to-be alma mater, Makerere University Business School. I tried not to think about the possibility that some of my classmates might see me standing there with my placard, and wonder what on Earth I was doing.

While Kampala is a big city, with 1.65 million people, it can sometimes feel like a small town, because so many Ugandans move there to look for work. So, the likelihood of seeing an alum from MUBS or my old school was fairly high. I tried not to think about that! Instead, I concentrated on the pedestrians and people in their cars and taxis. By now, our group of strikers was all getting pretty hungry and tired. When the alarm went off for the fourth time, we bundled up our placards and took a *matatu* home.

My mother was waiting for us, along with a full breakfast. She asked us how the strike had gone and what kinds of reactions we'd received. I mentioned the woman who'd spoken about the school and the trees being felled. To my surprise, Mom replied that she'd heard about that too, and suggested we might visit there sometime and strike to try to save the trees. I think she was proud of us. Now, every time she catches a news item on television about floods, droughts, or wildfires, she calls me to come and watch it.

"What have you been doing?" my father asked when he saw

we'd all returned. He hadn't known about the strike plan, and was genuinely intrigued by what had made the six of us get up and leave the house so early.

"We've been on a climate strike," I answered.

He looked at us in confusion, but not anger. Although my father had no idea what a climate strike was, I knew he could see the value in raising awareness of environmental issues that affected people's lives. For several years, he'd been involved in tree-planting projects with the Rotary Club, of which he was a leading member.

To my surprise, my dad said, "That's good." Then he urged us all to sit down and eat.

My sisters Joan (then aged 17) and Clare (19) were more inquisitive. "How did people look at you?" asked Joan. "Was it easy? Was it hard?"

Clare added quietly: "I think you were brave."

I was a little stunned that my sisters were interested. They hadn't expressed much interest in joining the strike when I'd confided in them the day before. They hadn't given a specific reason why they didn't want to come. Perhaps it was because it would be on a Sunday, and they wanted to sleep in, or like most people they didn't really understand what a climate strike was. Or maybe they didn't want to be seen by some of their friends or be in a photo that would be posted to social media.

As for me, I was elated: partly because demonstrating was new to me; partly because we young people had expressed ourselves and had tried to inform Kampalans about something that was affecting people in Uganda. I was also pleased for myself. I'd exposed myself to ridicule, but I hadn't let that stop me. I hadn't let my own worries overwhelm me, nor had I been put off by whether someone I knew might see me and judge me harshly. And I hadn't been so afraid of strangers criticizing me that I'd walked away from the strike.

Later that day, as I'd planned, I posted a few photos and videos taken by my brother Paul Christian to my 500 or so followers on social media. When I checked my phone again, I was happy to discover that some friends had liked my posts, and a few had even added supportive comments. Before I went to bed, I took another look at my phone. When I'd posted, I'd added hashtags, linking our climate strike to #FridaysFor Future. To my amazement, I saw that Greta Thunberg had retweeted the photos I'd uploaded, and suddenly my post had more than one thousand likes. My past posts had never received more than ten! *How had this happened?*

The next morning, I saw even more likes, and they were coming from people all over the world. Almost immediately, I began to plan my next strike for the following Friday. I'd started something. And I knew I couldn't stop.

2

Striking Out

As the days of the week ticked by, I became increasingly nervous, and also determined to see my first Fridays For Future strike through. This time, I was likely to be alone. My brothers had returned to boarding school, my cousins Nathan and Varak had headed home, and Isabella would be at her classes. I asked some of my friends to join me in the strike, but they didn't buy the idea. Like my sisters, they didn't understand what I'd be doing, or they weren't ready to let anyone take a photo of them holding a placard that could be shared on their WhatsApp friend groups—what we call the "walk of shame."

When Friday morning came, I ate breakfast, put on my jeans, and headed to my family's shop. I explained to my coworkers, who'd already seen the photos I'd posted from the first strike, that I would be striking again that morning. I asked them to cover for me if my father showed up, as he did once in a while, telling him I had an errand that required me to be out of the office.

I decided to revisit two of the locations from the first strike. I walked to Bugolobi Stage with my placard at my side. The Stage was full of people, and this time there could be no mistaking who was being stared at.

It's funny how your mind projects your fears onto others or replays the judgments you've picked up from society. Every time a car or taxi passed, I'd wonder whether a passenger recognized me or if they were thinking to themselves, *Instead of applying for a job, she's standing on the corner.* Or they'd be muttering under their breath, *Why isn't she helping people in poverty?* or *Why isn't she protesting for those who don't have enough food?* Or they'd be angry and asking themselves why I was wasting my time and theirs. I took every weird or quizzical glance, every wrinkling of the nose or narrowing of the eyes or frowning forehead, as a direct criticism of what I was doing.

I spent half an hour at Bugolobi Stage before I moved on to the Village Mall. To my relief, I was joined by my friend Elton John Sekandi. (Yes, he was named after a famous musician; and so, in a way, am I. In the mid-1990s my dad was a fan of the songs of American singer Vanessa Williams, and that's partly how I got my name!) Elton had told me he'd try to come if he could get out of work. Elton was from Masaka, about three and a half hours west of Kampala, also on the shores of Lake Victoria. I knew him because he'd been living in Kampala and had been working in some of my family's shops, including the one where I helped out.

Elton's presence made the second half of the strike a little easier, although the general indifference we encountered was disheartening. Because no one yelled at us or asked us a question or engaged with us at all really, it was hard to gauge how effective we were being. Was the message on my placard—a combination of the slogans we'd had in the first strike—confusing? Did people think the climate issue wasn't worth protesting about or drawing attention to? I had to keep reminding myself that only a few months before, I also hadn't recognized the gravity of the crisis.

One person *was* interested, and that was Elton himself. He asked me how I'd come up with the idea of striking for the climate, since he'd never seen such a thing before and didn't think it was possible. I told him about Greta and the Friday school strikes taking place in many countries. Our conversation must have had an impact, because he showed up at several of the later strikes I organized before he returned to Masaka to attend college.

In fact, encouraged by Elton's presence at that first Friday strike, I asked him whether he'd join me two days later, Sunday, January 13, for the next one. This time I suggested a different location: the Ugandan Parliament.

*　　*　　*

Striking outside Parliament, in the administrative center of Kampala, would make a stronger statement than the four commercial locations I'd protested at previously. By striking at malls and transit hubs, I'd been asking Ugandans to reflect on how our individual and collective choices as consumers affect the planet. But, as climate activists have made clear, the point of our strikes is not simply to raise awareness among other citizens, it's to push for ambitious, systemic change in government policies, private sector behavior, and investments. That means that not only must the drivers who put fuel in the tanks of their cars alter their habits, but oil and gas companies and all the other corporations responsible for greenhouse gas emissions must revise their business models and practices in favor of radical sustainability. For that to happen, there has to be economic *and political* pressure. Governments must listen to climate scientists; pass legislation, regulations, and budgets that decarbonize society and finance adaptation; and the biggest polluters must cut emissions to zero as soon as possible.

So, I decided I had to emulate Greta; I had to strike in front of Parliament.

This decision was potentially consequential. Elton and I knew we couldn't just walk up to the front of the Parliament building and hold our signs. The building itself was behind a gate and the entire surrounding area would be filled with security personnel. They wouldn't hesitate to ask why we were there and what we were doing. They'd also have the power to arrest us. I absolutely knew I couldn't tell my family about *this* strike, because they'd definitely see it as too dangerous.

But I was determined. That Sunday morning, I let my parents know roughly when I'd be back and took a *matatu* (public minibus) with Elton to a stop near the Parliament building. When we got to the gate, our stomachs hit the floor. We hadn't realized that that particular Sunday was the first day of Parliament Week, a series of events featuring parliamentarians and ministries organized, as the Government's website says, to "bring Parliament closer to the people." A charity walk to raise funds for Ugandans living with albinism was underway, so more people than usual were there, along with some political leaders. There was an enhanced police presence too.

"I'm scared," muttered Elton, as we entered the plaza.

I looked in the direction of one of the policemen. I thought it would be wise to tell the officers (truthfully) that our aim was to raise awareness about the climate crisis and wasn't a protest against the Government or the police themselves. I had to balance my attempt to communicate in front of my national parliament the urgency of the climate crisis and the need for governments to act, with the reality that however we might phrase our statements, they could be deliberately misconstrued or seen as a threat by the authorities.

"OK," I said. "I'll talk to him and explain why we're here, and see how he reacts."

I walked up to the officer, my heart climbing toward my mouth. People had been arrested for much less than carrying a few signs.

"Hello sir," I began hesitantly. "We're environmental activists, and we've brought along a placard to show our leaders why they should care about the climate crisis."

The officer frowned and his eyes narrowed.

"Have you been sent from the opposition disguised as climate activists?" he asked.

I assured him we hadn't.

He looked at me warily. "Are you targeting the Government to make it look bad?"

"No, we're not," I replied.

The officer, now joined by four others, read the messages listed on our placard:

GREEN LOVE GREEN PEACE

BEAT PLASTIC POLYTHENE POLLUTION

THANKS FOR THE GLOBAL WARMING
[that sarcastic one I still loved]

CLIMATE STRIKE NOW

I was aware the final line might bring up questions, and I'd debated whether to use the word *strike*, given its associations in

Uganda with unrest. I decided, though, it was important to make clear to those reading the sign that action was essential.

Sure enough, *strike* really disturbed the policemen, likely because the Makerere University tuition strikes and strikes by labor unions had often become violent and had been politicized by the Government and opposition parties. Ugandan politics is fractious, with the ruling party and the opposition often at loggerheads. Before or after elections, those conflicts can even be deadly.

"Are you students going to instigate strikes against the Government?"

I shook my head no.

"What kind of strike are you doing?" asked another officer sternly. "Are you sure you're not with the opposition?"

We told them again that we weren't, that our strike wasn't political in nature, and that it would be nonviolent.

Still, the officers resisted. They told us they wanted to be sure the placard didn't carry "hidden messages."

Finally, they relented and agreed that we could show our sign as long as we stood in a location they specified and called what we were doing an "awareness program." What a relief!

Elton was still nervous. I reassured him that if the officers had no problem with us, then we had less reason to be fearful. "If anyone asks us why we're here," I said to him, "we can say that the police allowed us."

Elton and I stayed at that designated spot for what turned into a ninety-minute strike. Given where we were standing, I've no doubt we were under surveillance, because the area around Parliament is full of cameras pointing in every direction. As it turned out, the officers neither threatened nor harassed us, and nor did anyone else. In fact, it was a hot day, and one of the policemen brought us a drink. Like Elton at the first Friday strike, two of the officers seemed intrigued. One told us he had a farm and needed some seeds. Did our organization provide them? We said we didn't. He then asked if he could become a member of our organization and, if he joined, would he be paid? We had to say no to that, too, of course. Not only didn't we have seeds, we didn't have an organization, but we didn't tell the officers that!

Before Elton and I left, we asked the policemen if they'd like to join us for a photo. Two of them let us take a picture with them, the others said no: clearly, they were anxious about what would happen to it. "Where will you post it?" one asked. "Who will you show it to?" the other inquired. We explained that we'd use the photos for social media, but the two officers still refused.

"Thank God this is done!" Elton exclaimed as we packed up to leave. We walked over to tell the policemen we were going, and to thank them for letting us hold the strike, or rather promote our "awareness program." I asked if we could do it again on that spot in the plaza. The officers said they weren't sure, because it depended on the day of the week and what else was happening.

I realized that as I talked with the policemen I'd been nervous but also strangely calm. I didn't want to appear anxious, because if I had I might have freaked out Elton even more. I felt an obligation to protect him. He was three years younger than I was, and had joined at my request.

To date, that has been my only climate strike in front of Parliament. When I returned by myself two weeks later on Friday, February 1, even though fewer policemen were present, I was turned away. When I showed the officers some of the photos we'd taken from the first strike, one of them replied that it had only been possible because it was on a Sunday. "Today," he said, "there are many people passing by, and what you are doing may instigate a protest from the public."

I realized I might need other approaches in trying to convince the Government. Five months later, in June, I composed a letter to Uganda's president Yoweri Museveni. I told him I was a young Ugandan "fighting for a better future for all." Climate change was "a threat to Uganda as a whole," I added. Then, I got to the heart of my message:

Climate change is real and dangerous for us all. It is happening right now. Scientists all over the world agree to this. The carbon levels are at the highest they have ever been. The climate crisis is mainly driven by human actions and it is increasing rapidly. We have experienced extreme weather conditions as a country starting with the very high temperatures in the month of January, followed by strong winds that caused destruction in various parts of the country. The heavy rains are causing floods, hence claiming lives of many people and leaving property destroyed. The climate crisis knows no age. It affects us all. Lives can be lost any time due to the impact of climate change. Africa as a continent is the least emitter of carbons but it will be the most affected by the crisis, including our dear country. This is why I believe that now is the time to take action and reverse the situation. We don't have enough time but I believe that if we start now, something can be done. As the President, I believe that you have all the authority and resources to help mitigate the crisis. Personally, I don't know if you will read this letter but I pray that you do.

In the end, I didn't send the letter. I was still learning how to be an activist and had little confidence what I'd written would arrive on the President's desk, let alone that he'd read it. Instead, I posted a picture of the typed letter and a video of me reading it to my social media pages. Honestly, it wasn't hugely popular there, but a few weeks later, Greenpeace Africa shared my post, which helped it reach a larger audience.

I wish it were possible to say that the situation has eased for activists who wish to alert the Ugandan people to the climate crisis. However, in February 2021, two years after my strike at the Parliament, three of my colleagues from the Rise Up Movement, which I founded not long after that first parliamentary strike, were arrested by twenty police officers outside the Parliament as they were conducting a climate strike. Even though what Evelyn Acham, Rebecca Abitimo, and Ayebale Paphrus were doing was no different from what Elton and I had done in January 2019, the authorities informed the three of them that they could never again hold a strike of any kind in a public place. Afterward, Ayebale told *Nile Post* reporter Jonah Kirabo: "One officer in command asked for our names, took pictures of us and told us to watch out because they could do anything to us and no one will ever know."[1]

* * *

Thursday January 17 was graduation day, and I couldn't have been happier. As the first child in the family to graduate, I knew completing my degree meant a lot to my parents. Still, I wasn't going to miss the opportunity while celebrating to also draw attention to global warming. I stood in my black graduation cap and gown, with shiny silver sandals on my feet, and the same sign with the four slogans that I'd had at Parliament in my hands. I posted the photo to my social media channels with the hashtags that by then I knew well: #climatechangeisreal, #fridayclimatestrike, and #FridaysForFuture.

From then on, I conducted strikes once or twice a week, slipping out of work each Friday morning and returning at around 10 a.m. I always posted photos from the strikes online, and I saw the number of followers and likes growing . . . slowly. I also began to receive interest from some climate organizations and media outlets. After my first strike, Climate Kids, a project of the US-based Climate Science Alliance, interviewed me. A few months later, the Ugandan *Observer* published a profile of me, above the fold on the front page. That was something my parents could relate to, and it made them realize how serious I was about climate activism, and that it hadn't merely been a project to tide me over until graduation. My mother told many of her friends to buy a copy of that day's newspaper, and some friends even posted photos of the front page on their social media. After the story appeared in the *Observer*, NBS TV, a popular national station, invited me on.

The message was beginning to get out through social media too. I was contacted by Green Campaign Africa in Kampala, who wanted me to talk to the young Ugandans they worked with about how they could join the climate movement. One of those students was Hilda Flavia Nakabuye, who became an FFF (Fridays For Future) Uganda leader and helps run its social media channels. Around that time, I also met Sadrach Nirere, founder of the #EndPlasticPollution Movement, which urges students and universities in Uganda to go plastic-free.

I also tried to encourage older students to become engaged in climate change activism and environmental protection. A few months after I graduated, I returned to Makerere University Business School with fourteen other activists. The semester was just starting, so many students were around. We dressed in green and collected trash and lots of plastic. Afterward, we conducted "person-to-person" climate education and awareness in front of the administration block, which is the hub of the campus. For about half an hour, we stood with our placards, which read GO GREEN BEFORE THE GREEN GOES, having conversations with the students who passed by. We followed up by attending a bazaar on campus where a DJ was playing music for the crowd. I asked him if I could use his microphone to address the students. I'd written a short speech, just in case I had an opportunity to give one. Those were my first public remarks on the climate crisis. My final words were, "Join us!"

"Join us." It's a message all activists for a cause promote and yet it's often difficult to find people to do that. Although spending time with Elton, Green Campaign Africa, Hilda, and Sadrach helped me realize that others were as worried as I was about the future for Uganda, in those early days of my climate activism, I was usually striking by myself.

This was a reminder that it's hard to be a climate activist in my country. Not only because the authorities can determine that any public gathering is political or violating laws and demand that it end or break it up. It's also difficult to coordinate a demonstration without a permit or an established organization to help you or provide you with an official stamp of approval.

For instance, during my first year of activism, I took part in a children's climate march in Kampala with thousands of participants, led by Little Hands Go Green, a national environmental organization founded in 2012 by lawyer and marketing professional Joseph Masembe that encourages children to plant fruit trees across the country. The Government supported marches like the ones organized by Little Hands Go Green because they weren't strikes or demonstrations. With mainstream credibility Little Hands could also receive backing from other organizations, as well as a number of schools.

Sometimes, the challenge is a matter of logistics and cost. When you're starting by yourself, nothing much is needed, but the more people who join you, the more you require. For the Global Day of Climate Action in September 2020, the Rise Up Movement was lucky enough to receive a donation that helped us make a banner, buy a megaphone, find a suitable strike location, purchase protective masks for the protesters amid the Covid pandemic, arrange transportation to the site, and provide drinks for participants when the strike ended. It wouldn't have been possible to do this without the donation, even if we'd had a permit.

Another difficulty was that many in my small circle of women friends, whom I'd met at university or boarding school, were skeptical about the strikes. Not only were they concerned about the potential social stigma associated with being a young woman holding a sign on the street, but, as twenty-two-or twenty-three-year-olds, they felt uncomfortable taking part in FFF activism, since FFF was founded and is mostly run by teenagers. My "ancient" contemporaries, including other recent university graduates, worried they didn't belong or that it would have felt regressive to join in.

Even though I could understand their hesitation and I wasn't a student anymore either, I didn't mind being associated with school-age protesters, and I didn't care about the day of the week on which the strikes were held. After all, in the case of my first strikes in Kitintale and Bugolobi, my brothers and cousins were school-aged, and it was a Sunday. I was also motivated by these young people in a number of countries, many of them women, stepping up to fill the gap left by older generations who had failed to prevent the climate emergency.

However, in order to overcome my peers' resistance, I suggested we form a group and call it Youth for Future. This would distinguish us from FFF, but we'd still support FFF and work alongside it. This made some of my friends more comfortable, and encouraged them to hold strikes from their homes, posting photos on social media of themselves holding signs, or on occasion to join me at Bugolobi Stage and elsewhere with their placards. By the turn of 2020, Youth for Future had morphed into the Rise Up Movement, which I thought could better encompass activists beyond Uganda. You'll learn about the work of many of those activists in this book.

Nonetheless, whatever name you call a group, gendered ideas in my country about what roles women should play inhibit more of them from joining strikes. Once a woman graduates from college, it's presumed she'll get married soon and, not long after that, give birth to her first child. If she's been "raised well," she'll be expected to dedicate herself to her kids and to cooking and cleaning. Even if she has a job outside the home, society reinforces the assumption that childcare and household chores are her responsibility. Not many men challenge that division of labor.

Finally, the concept of the strike remained a hurdle, and not just to police officers outside the Ugandan Parliament. I wanted to use the word *strike* to connect to the FFF strikes, which were gaining worldwide attention. I loved the fact that FFF activists were seeking to take control of their future and demanding to be heard. They were also asking fundamental questions about education: If adults weren't following the facts and not telling the truth about the climate crisis, what was the point of education in the first place? If knowledge didn't lead to action, and the future that education was meant to prepare you for was being mortgaged by the same people who were telling you to go to school, then why not take to the streets to ensure there *was* a future rather than sit in a classroom?

While I fully understand how such a stance is possible in countries where public education is free for all children, it's not possible in Uganda and most other African nations for students to leave their classrooms or not show up at school at all (which is customary in an FFF strike). A student who walks out or misses lessons, unless they have a family or health emergency or another compelling reason to be absent, risks suspension or expulsion. Almost all schools beyond primary (elementary) level in Africa are fee-paying, and there are no tuition refunds for students who've been expelled. (At least, I've never heard of such a case.) In Uganda, we also have fees for primary school, with a few exceptions. For parents who've paid to send their child to school, and whose welfare may depend on that child completing their education and getting a good job, a son or daughter skipping school can be seen as a betrayal of the whole family.

In addition, many students in Uganda attend boarding schools, especially for secondary (high) school as I did, and they're often located, like mine was, in the countryside. While you could theoretically stage a strike there, you wouldn't have a very big audience—and you'd still be risking punishment or expulsion.

So, since I'm always looking for solutions that reflect reality and the need to get the message out, I decided that instead of suggesting that students walk out of classes, I'd try to take the climate strikes *into* schools—where they could form part of the curriculum in a way that I'd wished climate change had been when I was a young girl.

In March 2019, I visited Reverend John Foundation Primary School in Kampala. I explained to the principal I was trying to create climate awareness among students and demand that leaders act. She agreed I could speak to the pupils and stage a strike with them too. I was excited by the school's openness to collaboration. When I returned a week later, the teachers had assembled about a hundred students in the compound. I explained to the kids I was fighting for the protection of trees and the whole of our planet; protesting against single-use plastic that's made from fossil fuels; and also that I was trying to prevent more people from being flooded out of their homes or swept away in landslides. I kept the language simple and non-technical, and even led the students in a chant: "What do we want? *Climate justice.* When do we want it? *Now.*"

The teachers didn't appear concerned that what I'd said was radical or subversive, and even encouraged the students to chant louder! This became a model for the many climate awareness strikes I've organized in schools.

*　　*　　*

At my graduation lunch, many of my aunties made clear they were looking forward to the next big family celebration: my wedding. I find it weird that your family doesn't want you to date when you're in secondary school or at university, and yet when you've graduated, they expect a spouse to magically appear—just like that. The way your entire life is mapped out as

a woman is very annoying, and I try not to give it much attention. And I definitely don't let it distract me from my activism.

That said, society's narrow expectations for women in their
early- and mid-twenties are a reality. I've already revealed that I
still live with my parents: in Uganda it would not be socially
acceptable for me to live by myself or with someone I wasn't
married to. I could live with my sisters, but it's very unusual for
young single women to rent an apartment together.

The craziness of social norms extends, as I've said, to protesting itself. My dad told me that he'd heard complaints from
some of our extended family once I'd become a regular climate
striker. "What is she doing?" they'd asked him, disapprovingly.
"She's a graduate and I found her in town on the road holding
placards! What is wrong with your daughter?"

My dad stood up for me. "No, no, no," he told me he'd said.
"Let her go on." Then he added, by way of explanation (and
maybe mostly for my benefit): "When Vanessa decides to do
something, she really goes for it."

I'm fortunate my parents have sought to understand and
support my activism, and haven't pressured me to marry and
start a family. That openness may be a result of their upbring-
ing. My father, Paul, became an entrepreneur in his teens, and
has been interested in public service since he was young. He
was elected a youth leader in his community in 1998, when he
was twenty-five, and then became a town councilor for two
terms. In 2016, he ran to be mayor of Nakawa Division. Nakawa is one of the five divisions that make up the city of Kampala.
Each division has an elected mayor, who serves under the Lord
Mayor. My father lost that election, but he ran again in January
2021 and won.

My father has always been what I'd call a "people person."

When I was young, I remember people coming over to our house, and my mother would prepare breakfast, perhaps tea and bread, and he'd listen to their concerns. Even though I don't like politics myself—there's something about politicians I distrust—I've been inspired by his commitment to help his constituents and neighbors, and his participation in the Rotary Club of Bugolobi. I remember my father receiving multiple certificates from the Rotarians, whose motto is "Service Above Self." When I asked him why he had been awarded all his certificates, he told me he always puts his best efforts into whatever he undertakes, especially when it comes to assisting others.

My mother, Anna, is also passionate about helping others. She's generous with her time, often assisting neighbors and others in the community by sharing food, helping secure health services, and offering advice. She's a twin, and she traces her origins to a village in Kagoma District, in northern Kampala. From what she's told me, her childhood wasn't easy. She recalls having to run a lot, and her family always kept their bags packed, a consequence of the conflicts that have convulsed Uganda and neighboring countries over several decades. Sometimes all she and her siblings had to eat was porridge.

She also saved my life. When I was a small child, Mom and I were walking to the market one morning. It had rained heavily overnight and the water had washed away some of the soil on the roadside. As we walked along, I stopped and pulled back, and when my mother looked down to see why I was resisting, she saw me bending down about to pick something up. It was an unexploded bomb.

Immediately, my mother pulled me away and called for assistance. A military officer confirmed it was an explosive device, and evacuated the immediate area. Soldiers from a nearby barracks came and removed the bomb. My mother was told that it was likely that the Allied Democratic Forces (ADF), an insurgency group that has operated in Uganda and the Democratic Republic of Congo for more than twenty years, had planted it. They had probably placed it at the roadside to blow up a car and kill many people in the subsequent explosion.

My mother doesn't talk about the incident much, but she and it are reminders to me of what matters: that we need to take every opportunity to live a meaningful life, because you never know what can happen.

*   *   *

Looking back two years on from those early days, I'm both surprised and not surprised that I became an activist. Even though I'm shy and love to be by myself, I'd served in leadership positions. I was an assistant house captain at boarding school, responsible for sports, music, dance, and drama, and later I became prefect for academics and even participated in "Miss Teen," a national youth talent competition, referred to mistakenly by some local media as a "beauty pageant." Participating in the competition was an important turning point for me. I had just started at upper secondary, the final two years of high school, and it was taking me some time to adjust to everything being new. Even though some students were critical of my entering and told me I could end up embarrassing the school because I was so reserved, my heart wanted me to do it. But I shocked everyone, including myself, when I was the second runner-up for my school.

Being in the contest was the first time I had to speak and answer questions in public, and it gave me some confidence, which made my remaining years at the school much easier than they might have been. My positive experience with Miss Teen is perhaps why I haven't minded too much standing up in front of an audience and speaking, although, like many young women, I'm still ambivalent about it. I'm not immune to the messages many young girls and women receive that we should be quiet and not put ourselves forward as spokespeople or decision-makers. We're encouraged or expected to cede our voices and authority to boys or men. Too many of us listen to those voices inside us, and those that might come from our peer group, that tell us we shouldn't speak out or stand up for what we believe in.

During that early part of 2019, even though I had many moments of doubt, and worse, my activism grew—along with my passion, conviction, and purpose. Continuing my weekly climate strikes, doing the campus cleanups and in-school climate education, coordinating with other Ugandan climate and environmental activists, writing, continuing with social media, and giving interviews made me feel I was moving forward.

I had to believe that some people were listening.

3

COP Out

Uganda's environmental catastrophes continued throughout my first year as an activist. In June 2019, Kampala was hit by flash floods, during which eight people died,[1] and Kasese in the west of the country was engulfed by rising water levels, burst riverbanks, and severe flooding, as it had been seven months earlier, when roads and hospitals had been destroyed and crops washed away. Bududa, which had suffered so much in 2018, was struck again by devastating landslides, as was nearby Sironko.[2] These events served as a constant reminder to me of why I'd become an activist.

Early in September 2019, I received an email from a woman working on climate issues in the office of the UN Secretary General. I was astonished. She asked if we could speak on the phone, and when we did, she told me that she and her colleagues had been following my activism on social media and they hoped I could attend the Youth Climate Summit at the United Nations headquarters in New York on September 21. They told me that the summit was scheduled two days before the Secretary General's Climate Action Summit, which was organized so countries would commit to bigger cuts in their greenhouse gas (GHG) emissions than what they'd agreed to in Paris in 2015. The UN would cover my plane ticket and my accommodations in New York, and would email all the information I'd need.

For nearly nine months, I'd been posting my strikes and climate education activities online. Even though I was getting more retweets and comments, I was still amazed that someone from the UN was paying attention. When I told my mother about the invitation, she thought it couldn't be true. My father was also speechless . . . and skeptical. He asked to see the email to determine if I was being pranked or scammed. "This is real," he exclaimed to my mother and me. "An authentic invitation from the UN!" This represented another milestone in my activist journey. My parents could see that what I was doing wasn't solely a personal commitment or something of only local interest or relevance. So, I accepted the invitation.

I had only a couple of weeks to get ready. Of all countries to travel to, the US presents the most difficulties for a Ugandan, and it usually requires months of preparation on the applicant's behalf, as well as an interview, to secure a visa. My father started calling around for advice from his friends on how to apply. Thankfully, I got the visa in time. In the days before my trip, my dad made me a business card, lent me his suitcase, and gave me $150 USD, "just in case."

I was both nervous and excited. I'd never gone beyond Uganda's borders, and I'd be the first in my family to visit the US. It would be my first time traveling by myself, or on a plane, or to another country, let alone another continent.

*   *   *

On September 18, as we threaded our way to Entebbe Airport through Kampala's evening traffic, I felt a mixture of excitement and nerves. As I boarded, it was all new: the seat numbering, the size of the plane, the thrust of the aircraft as it took off. I tried not to think about crashes or how far below me the ground was. It was hard to sleep, but finally, sheer exhaustion caused me to doze off.

When I arrived at JFK Airport in New York City, after my connecting flight from Amsterdam, I had to figure out how to get to the West Side YMCA in Manhattan, where I was to stay. There seemed to be a bewildering number of options. I was so discombobulated that I stood outside the busy terminal with my luggage thinking about what to do for nearly ninety minutes on that hot afternoon. Here I was: newly arrived in a strange country in an enormous city, with only $150 in my pocket and no credit card. No one was there to meet me, I knew no one else attending the conference, and I didn't see any other young people or groups of delegates who might be participating at either of the summits.

Finally, and even though I realized it would be the most expensive and least climate-friendly choice, I spent $70 on a taxi to the YMCA. The density and craziness of the traffic reminded me of Kampala. Once I arrived at the Y, a woman who worked with the UN was there to help youth delegates check in. I was assigned a room with another young climate activist, who wasn't there yet, and I didn't see any other delegates either. Once I got to my room, I promptly lay on the bed and fell asleep. Later, I heard a knock on the door, and I met my roommate, Amelia "Lia" Tuifua, a Fijian activist with global climate network 350.org. Lia said she was going to get some food with other Fijian climate activists, and asked if I wanted to join them. But I was so intimidated and exhausted by my travels that I stayed in the room, and Lia, my new friend and angel, brought me back some pizza to eat.

The next morning, a Friday, Lia and other Fijian activists told me they were visiting their consulate to seek help with additional expenses. Since the UN representative I'd met at the hotel the night before hadn't mentioned covering costs for food or local transportation during the days I'd spend in New York City—and given that I'd had to work out how to get to the hotel myself—I was realizing I'd have to figure things out on my own. Therefore, like Lia, I decided to visit my consulate and see if someone could assist me.

When I got there and explained my situation to the receptionist, she said they could only support the official government delegates to the Secretary General's climate summit. But then she very kindly gave me $50 of her own money, which was a lifesaver. New York is an expensive city.

I'd learned that US climate activists had arranged a march starting at noon, to apply pressure on leaders in advance of the UN summit. It would be one of more than 4,500 climate strikes in 120 countries that day. Organizers anticipated tens of thousands of people at the New York rally, which would begin in Foley Square in Lower Manhattan and end at Battery Park on the tip of the island. Greta Thunberg and other young climate activists would speak.

I was uncertain about joining the march, not least because I had no idea how far Battery Park was from the hotel, and there wasn't an organized meetup for the Youth Summit delegates. But I was excited too. I'd never participated in such a large protest—let alone one where everyone was calling for the same thing: *Climate Action Now!* When I arrived, I stood holding my placard YOUTH STRIKE FOR CLIMATE surrounded by tens of thousands of people, young and old. I thought about how different this experience was from standing at Bugolobi Stage or outside Parliament with Elton, or all the solitary climate strikes I'd done back in Uganda.

I was happy to be part of a climate action that had taken over the streets, which would have been impossible in Uganda. I was also thrilled to catch a glimpse of activists I'd been following online for months, like Alexandria Villaseñor of the US and Xiye Bastida from Mexico. And as we began to move and chant, I was drawn in and my anxiety lessened.

Yet, at the same time, I felt alone. I imagined the other youth activists attending the summit the next day might be there, but I couldn't see them. I'd hoped I might spot some fellow activists from Uganda, but I never did. I'd also anticipated I might be able to say hello to Alexandria to tell her how inspired I'd been by what she'd done in helping to organize the march. But the person guarding the barricades around the stage at Battery Park told me she was too busy. I was disappointed, but I could understand.

By the time I'd listened to many of the speeches, it was late afternoon and jet lag and exhaustion were kicking in. So, I found a bus uptown, ate at a deli near the hotel, took a shower, posted some photos to social media, and went to bed. It had been a long day, and (I reflected) I'd mostly experienced it by myself.

The next day, Lia and I went to the UN for the Youth Climate Summit. As we waited for it to begin, I looked around and noticed I was one of relatively few Africans or Black people in the room. I also had the strange experience of being asked to give up my seat twice because it was reserved for someone else, even though the seats didn't have a sign saying as much and I was the only one asked to move in that section. I had to stand for a while before I finally got a seat at the back.

Another disappointment was that, when I'd first been invited to the summit, I'd been told I could give a speech about the climate crisis in Uganda, but later I learned I wouldn't have a formal speaking role after all. Still, I was very happy to listen to Greta and teenage US activist Jamie Margolin, co-executive director of Zero Hour, a youth-led organization seeking to center the voices of diverse young people around environmental justice, which had helped conceive of and organize the climate march. After lunch, which the UN provided, I joined about thirty participants in a breakout session, and I was honored to be one of those presenting a summary to the Youth Summit delegates of our discussions on what our leaders should be doing.

A more promising moment in New York was when I met members of the Ugandan delegation to the Secretary General's climate summit. One of them had sent me an email while I was in New York, asking me if I'd like to come to breakfast at Uganda House, where the permanent mission to the UN is located.

I could barely believe my luck. Just a few months earlier, the police had turned me away from striking in front of Parliament, and now I'd have a chance to communicate directly with those in authority. At Uganda House, I was introduced to several members of the delegation, including the chair of the climate committee, who gave me his business card, and a Member of Parliament who said he'd seen me on *Face Off*, a public-affairs show on NBS TV back home. I discussed the consequences of climate change for Uganda with them. For someone on a minimal budget like me, the buffet of foods that reminded me of home—including what we call "Irish" (potatoes), tea, and bread—had never tasted so good.

I thought my encounter with government officials at Uganda House might be an opening—a chance when I got home to speak to the Parliamentary Climate Committee and even, perhaps, to the full Parliament itself. Unfortunately, when I returned to Uganda, the Climate Committee chairman never answered my calls, and the delegate who'd urged me to stay involved, and with whom I'd exchanged business cards, didn't respond to my efforts to reach him. Maybe he'd lost my card.

*   *   *

On September 23, the day after my breakfast at Uganda House, I attended the UN Climate Action Summit, along with other youth climate activists. Most people now remember that Monday for the trivial semi-encounter between a frowning Greta Thunberg and an oblivious Donald Trump. What I recall, though, was a collection of uninspiring remarks from prime ministers and presidents that failed to acknowledge the dire urgency of what Earth was facing. Of course, I also remember Greta's mesmerizing speech, which surprised government delegates and the media with its force and emotional honesty. It was startling to sit in that vast General Assembly Hall, which was so quiet and filled with so much intensity that when Greta finished speaking the applause seemed to literally shake the room.

The next day, I shared a taxi (and the fare!) to the airport with Lia and two of the Fijian activists. The long flights home gave me plenty of time to process my time in New York. I'd been incredibly fortunate. I'd seen several of my heroes, made some new friends, and been part of a massive climate march. I'd attended two summits at the United Nations, had met and eaten breakfast with Members of Parliament at Uganda House, and had spent time in a city and country that many people around the world only dream of visiting. What I'd enjoyed the most was that I was finally able to talk one-on-one with other activists. Being in their company had given me a tangible feeling that I was part of something truly global.

Yet I couldn't shake off my unease. As one of the few African activists at the Youth Climate Summit, I wished I'd had a chance to speak about our continent's realities. I'd hoped to learn firsthand from experienced activists and my climate heroes, instead of seeing them across a conference room or behind barricades. I'd felt lonely and isolated even amid the city's crowds and the climate marchers, and the pressure of keeping to the tiny budget I had in that strange, enormous city by myself weighed on me.

Perhaps I'd overlooked the email with the instructions on how to get to the hotel, and perhaps I could have been more forceful and asked the Secretary General's climate office about funds to cover my costs for meals and local transportation. Perhaps I could have contacted my fellow activists for assistance and advice. And perhaps my being asked to move twice was merely a misunderstanding. But all of it was new and overwhelming, and my reticence and self-containment held me back. Maybe that, too, was a mistake.

In any event, my family was at Entebbe Airport to greet me. It felt good to be back. Although my time in New York hadn't dampened my passion for, or commitment to, activism, my disappointment was so great that I didn't talk to anyone in my family or my friends about the trip. It might be hard to understand, but in some ways those few days in New York were heartbreaking.

*   *   *

My heartbreak, though, paled in comparison with what was happening in Uganda. There was no letup in the torrential rain that continued to soak the country. That October (2019), massive flooding in the eastern region left hundreds homeless, and the district of Kasese in the west was inundated once again.[3] Unprecedented rains in November caused people to be washed away when a bridge was knocked out in the central region.[4]

Each lethal flood or fatal landslide sharpened my sense of how much had still to be done and in such a short timescale for these terrible events not to become *much* more catastrophic and frequent. This was why the UN COP conference was so important.

Let me explain.

In 1992, following the United Nations Conference on Environment and Development, held in Rio de Janeiro and known as the "Earth Summit," the UN had established the UN Framework Convention on Climate Change (UNFCCC), which since 1995 had convened an annual meeting of the Conference of the Parties (that is, governments), known as the COP, to register progress or otherwise grapple with the most urgent issues related to climate change. COP meetings are hosted by a different country each time. The twenty-fifth climate COP would be in Madrid, Spain, for two weeks during the first half of December 2019.

Following COP 21 in Paris, all governments had made commitments through nationally determined contributions (NDCs), on how far and by which date they'd lower their emissions. These NDCs were to be assessed every five years. For Fridays For Future activists such as myself, the NDCs were wholly inadequate if we were to stand a chance of ensuring a ground-level temperature increase of less than an average of 2°C (3.6°F). Furthermore, countries in the Global South had anticipated that, after COP 21, the industrialized world would finally recognize that Pacific nations were being engulfed by rising seas and African nations were facing widespread food insecurity as a result of global heating. We'd expected them to cut their emissions quickly (not set vague future targets) and provide sufficient climate finance for poorer countries. These commitments hadn't been kept. So, COP 25 in Madrid would provide an opportunity for climate activists, especially from the Global South, to insist that those pledges be honored.

I wanted to be there. Fortunately, I was in a WhatsApp group with activists from Greenpeace, the international environmental campaigning organization, where we'd share updates on our strikes and other awareness and advocacy efforts. Greenpeace offered to sponsor some young activists to attend COP 25, and I was one of the lucky ones they said they'd support. (I also ended up being invited by Avaaz—a US-based nonprofit that supports young activists working on climate change, human rights, and animal rights—and by 350.org, a US-based global network seeking to end the age of fossil fuels and accelerate the transition to 100 percent renewable energy. I didn't want to offend any group, and Greenpeace invited me first so I decided to accept their sponsorship.)

COP 25 was a whirlwind. My flight to Madrid was delayed, so when I arrived, I had no time to change my clothes or go to the hotel. Instead, Jessica Miller from Greenpeace rushed me straight to a press conference with other activists at the conference center where the COP was being held. To my surprise, Greta was there too, and then, to my amazement, she turned to me and said, "Hi." A little later, Jess and I came across Greta, her dad, and a few other activists at the conference center. Greta and her father were sitting at a table eating various kinds of fruit, which they invited me to share with them. I was so hungry!

They were very welcoming and asked me about climate activism in Uganda. I told them it was hard to stage strikes like the ones in Europe and the US, and about how large-scale protests were impossible because our placards could be confiscated, and we could be arrested. I explained that that was why I'd been conducting strikes on school grounds and taking other actions. I told them how floods, heatwaves, and drought were destroying the crops on which the vast majority of Uganda's people depend for their livelihoods, and that food insecurity was a huge issue in Uganda. The resulting poverty, made worse by the climate emergency, meant that too many girls weren't in school because their families couldn't afford the fees. Others were being married very young so their parents could receive some food or money in exchange, which could be crucial to their survival, but disastrous for the girls themselves.

Greta and her father listened very carefully and told me how much they appreciated what I was doing. I was very grateful for that. There was no self-aggrandizement on Greta's part, no attempt to claim she was more important than anyone else. She was completely focused on pushing governments and businesses to follow the climate science and act with the urgency and at the scale required. I must admit it felt surreal to talk with Greta and her dad about *my* activism when I had so much admiration for hers.

When I finally arrived at the hotel, I discovered I'd be sharing a room with eight other youth climate activists from places like Russia, Chile, and Spain. I relished the chance to learn about their activism and share ideas and strategies. Unlike my visit to New York, Jess and the Greenpeace team kept me and the other activists they were supporting updated about what was taking place and where. That was really important since the COP has a number of parallel venues, for events organized by non-governmental organizations (NGOs), activists, and universities, among others.

Greenpeace also introduced me to other activists from the Global South (we were still outnumbered by those from the Global North). I met Licypriya Kangujam from India, who'd only just turned eight, and was calling for India to reduce its high air-pollution levels. At the COP, Licypriya urged leaders to pursue an issue I also believe in passionately: that climate literacy should be taught in schools. I also hit it off with three activists from the Cayman Islands in the Caribbean, Olivia Zimmer, Steff Mcdermot, and Connor Childs, whose approach really inspired me. They were using art and photography to spread climate change messages online, which was something I hadn't encountered before.

Another highlight of Madrid was connecting with other African activists—such as Hindou Oumarou Ibrahim, who's the president of the Association for Indigenous Women and Peoples of Chad. She campaigns to bring more Indigenous People—who have so much wisdom about climate adaptation and genuine sustainability—to climate change deliberations like the COP. Hindou is an advocate for the UN Sustainable Development Goals (SDGs), and I'm honored now to join her in that effort. (I write about the SDGs and my involvement in Chapter 9.)

I also linked up with activists from my own country, three of whom are my friends and climate colleagues. There was Leah Namugerwa, who was then fifteen and who'd been trying to enforce a ban on plastic bags in Uganda; Davis Reuben Seka-mwa, who'd taken part in a MUBS campus cleanup and had carried on the Friday climate strike in Uganda when I'd been away in New York; and Hilda Nakabuye, whom I'd met earlier in the year.

When I wasn't meeting or strategizing at the COP, I was being interviewed by a number of journalists, which gave me an opportunity to highlight how climate change was already severely disrupting lives, livelihoods, and ecosystems in Uganda and across Africa.

I especially appreciated the unexpected opportunity to speak at the Social Summit for Climate Action, a weeklong parallel gathering to COP 25, organized by environmental and social justice activists. I was sitting in the audience with Davis and Sil-via Díaz Pérez (an organizer from Greenpeace), waiting for the talks to start, when one of the organizers asked me if I would speak, since they wanted someone from FFF (Fridays For Future) to present.

"But I'm not prepared," I said to the organizer. "What will I say?"

"Just do it," urged Silvia. "You'll find the words."

"Speak what comes to your heart," Davis added.

Eventually, I agreed, and I ended up talking for longer than I thought I would. The words poured out: my strikes, the epidemic of child marriage, and the other terrible effects of the climate crisis in Uganda. I reminded fellow activists what brought us to the COP and urged them not to accept the green-washing of the big polluters and the leaders of those countries that supported them. We had to hold them accountable, I said, and speak out no matter what happens. I was truly grateful for the opportunity to talk, even though it was entirely unexpected. I was also happy that what I said seemed to resonate with many people in the room.

I also took part in a number of strikes at the COP. During one, we occupied a stage, our hands raised, and chanted for climate justice, until security officers forced us to leave. At another, more than a hundred of us joined with Greenpeace and FFF activists in a sit-down in the lobby of the building where the governments were holding their negotiations. We chanted until conference security guards forced us to leave and banned us from the COP venue for the day. This infuriated us, because corporations—some of which were major greenhouse gas emitters—were still in the conference hall.

Some of you may feel uncomfortable about sit-ins and strikes, especially when they take place in the presence of government officials engaged in high-level discussions. I understand that unease. Throughout 2019, I wavered between my wish to speak out to make people pay attention and my natural shyness. I struggled with the discomfort of being defined as disruptive and my fear of what authorities might do if they judged my activism to be illegal. I think for me that uncertainty will always be there. To be with those activists at COP, chanting and occupying the lobby, however, was exhilarating. In Uganda, we never could have done something like that. I can imagine we might have been harassed, possibly tear-gassed or beaten up, or even arrested.

Now, sometimes there will be situations that are too

dangerous or volatile for such a protest. But it's my firm opinion that the time for deference and patience on the climate crisis is over. We can no longer assume that a well-meaning attitude or a slow adjustment of policies or a slight shift in production is sufficient. This is especially true if you're my age or younger. Leaders aren't making the decisions we need, the planet is being destroyed, people's livelihoods are being wrecked, and hundreds of thousands are dying: we *have* to protest. We can't let leaders believe that they can do what they want because there aren't enough people who care or who aren't bothered about the wrong direction they're taking us in.

To anyone who thinks youth climate activists are too immature or naive to raise their voices on behalf of others, I say that maturity is not about your calendar age. You may be as young as Licypriya, born in 2011, or as old as the broadcaster and environmentalist Sir David Attenborough, born in 1926. Both of them love this planet, have listened to the scientists, and are trying to convey how high the stakes are, to the best of their abilities.

I absorbed an enormous amount at COP 25. Through meeting so many FFF activists from different countries, I learned more about how to coordinate strikes and protests and to keep them going. I became more confident speaking at press conferences and less intimidated by journalists, especially when there were so many. I was also added to a number of different FFF WhatsApp groups, which has proven very useful for circulating news and updates, alerting people about arrests, and maintaining solidarity across borders.

What I learned at COP 25 would help me sustain and expand my activism and education projects when I returned home in mid-December. As I left Madrid, I resolved to apply more pressure to the fossil fuels industry and the forces speeding deforestation. I also thought more about issues of representation that I'd reflected on in New York. While there were a number of activists from the Global South at COP 25, there were many more from the Global North, and it struck me there should be parity at all future climate conferences.

*　　*　　*

When I think about my first year as an activist, certain realities become clear to me. I've come to appreciate that even though you may be protesting alone, you cannot anticipate your effect on others, whether they're in front of you or online. I found out I truly was part of a global movement and that when there is unity and coordination, we can do more and have more of an impact.

I discovered firsthand how valuable it is to meet together (which, of course, the pandemic made impossible), because human connection and shared space are irreplaceable. I saw how people of all ages care passionately about the climate crisis and are confronting the same frustrations and challenges I am. And I understood how important it was to learn from them, be in solidarity with them, and be inspired by them. Their struggles were my struggles; my victories were theirs, and theirs were mine.

But I also learned another valuable lesson that year—one that many of us find hard to talk about. By the middle of March 2019, I had been conducting weekly climate strikes for more than two months. While I believed in the importance of maintaining the strikes, it was dispiriting that others only joined me on rare occasions and that so few people in Kampala seemed to be engaging with the messages on my placards. The traffic never stopped on Jinja Road; cars and taxis pulled in and out of the Shell station; the shopping carried on at the Village Mall; and the planet continued to burn. The more I discovered how many ways people were being affected in Uganda and elsewhere by the climate crisis, the more personal the strikes became for me. And the more this happened, the more profoundly it hurt me that almost no one in my country seemed to be responding to the larger emergency. I fell into a depression.

By April, I was feeling very low, and for two weeks I couldn't summon the energy to strike. I remember crying a lot alone in my room, because nothing was making sense. I announced online I was taking a break from climate activism for a while and logged off social media. I didn't have that many followers then, and I don't remember many messages urging me to reconsider.

Later, I learned that what I was experiencing—burnout—isn't uncommon for activists, especially ones new to a cause. Like them, I had discovered something that disturbed me. For the climate crisis, it's the sheer scale of the problem. I can't understand why more people aren't outraged or that more institutions, organizations, and governments aren't mobilizing—and that's also disturbing. I stood in the street week after week, and nothing appeared to change. Even our own family members or friends may not understand why we care so passionately about something that a few months earlier we hadn't a clue about. Even if we find support online or are part of an activist group, we can feel isolated, ignored, and useless.

I can see now that my tendency to hide my feelings wasn't helping me. Somehow, I didn't believe my family would understand. And even though I admired the many international climate activists who appeared so strong and committed, I didn't think I could reach out to them and confess how weak and defeated I felt. I wasn't even comfortable talking about my feelings with the environmental activists in Kampala I'd become acquainted with. It also felt too daunting to explain my situation to long-time friends from school and university, even though in the past I'd confided in them when schoolwork or social pressures had become too much. Would they accept that I could be depressed about something as abstract and impersonal (in their eyes) as the climate crisis?

Around this time, I realized I *did* have a friend, also an activist, to whom I could and did finally talk to. Davis Reuben Sekamwa was, like me, at the stage of trying to find ways he could make an impact as a climate striker in Kampala, in Uganda, and beyond our borders. Davis gave me some very important advice. Self-care, he said, was essential. If I needed to take a day off, then I should. "The most important thing," he told me, "is not to give up on the vision you had when you started your activism."

It took a bit of time, but I grew to see that Davis was right. I'd only been an activist for a few months, and I was judging myself to be a failure. My expectations were self-defeating and counterproductive. I needed to remember, as Davis had said, that self-care wasn't an indulgence but a necessity. He reminded me that the work was too important for me to burn out. I restarted my strikes and, after about a month, I rejoined social media.

As any activist who's ever suffered such lows will recognize, it's not easy to shake off feelings of inadequacy, doubt, or despair. That's why you have to develop a community, whether online or in person, of people who support you and whom you can support. And after attending COP 25, I felt I'd found more of that community.

*   *   *

Activists weren't surprised that the government deliberations at COP 25 ended in a massive disappointment. Major issues were left unresolved. Wealthier countries again refused to fulfill commitments they'd made *in 2009* to marshal US$100 billion a year to support poor countries' adaptation to the effects of climate change and to reduce their GHG emissions. Another disappointment was their unwillingness to cover the costs of the "loss and damage" caused by the climate crisis. The final text the government negotiators agreed to in Madrid admitted that the emissions reductions governments had already pledged fell far short of those necessary to meet, let alone exceed, the 1.5°C and 2°C (2.5°F and 3.6°F) temperature increase targets that had been set at COP 21. The language governments used was weak, conditional, and completely inadequate to the unfolding catastrophe.

How could I explain this outcome to other Ugandans, for whom the climate crisis was literally a matter of life and death? As I celebrated Christmas and New Year at home, I resolved to try to expand my activism in Uganda in 2020 and organize with other activists to bring more African voices, perspectives, and solutions to international climate policy-making.

4

Crop Out

E arly in January 2020, Arctic Basecamp, an organization of scientists raising the alarm about the rapid warming of the Arctic, invited me to Davos. I was intrigued. I would join five other young activists and sleep in a chilly expedition tent outside the venue to show attendees at the World Economic Forum that they, like us, had to move beyond their comfort zone if we were to address the climate crisis.

At that point, I'd never heard of Arctic Basecamp or Davos or, for that matter, the World Economic Forum (WEF). When I told a friend who's also now an activist that I'd accepted the invitation, she said she'd researched the event and that many rich people would be attending. That didn't interest me as much as the prospect of seeing snow for the first time. I knew that Switzerland was a mountainous country, and since it was January in the northern hemisphere, I assumed it would be cold. Arctic Basecamp's team told me they'd be supplying the winter gear I'd require, so I needn't worry.

I once again borrowed my father's suitcase, which was heavy and had wheels, and packed it with all the layers I thought I could wear. Then I set off for Entebbe Airport, flew to Zurich, and got my first glimpse of the Alps. I took a train to the ski resort of Davos and arrived around midnight.

When I stepped out of the heated carriage, the cold hit me like a sledgehammer. In my effort to make sure I caught the correct train in Zurich, I'd forgotten to remove my gloves from my suitcase and I was wearing just a sweater and scarf over my shirt. I could see only a few streetlights, and I grew nervous. For a woman in Uganda to be outside alone at this late hour and on a badly lit street would be considered reckless. Plus, I had no idea where to find the funicular railway to take me up the mountain to where the Arctic Basecamp tents were set up. Thankfully, I came across an old man who noticed how lost I looked and he asked some teenage boys to help me drag my suitcase to the funicular station.

By this time, my hands were aching from the cold, and everything I touched hurt. My body was shaking to keep warm, and I'd begun reciting prayers in my head. What I hadn't grasped was that the funicular departed only every half-hour at that time of night, so I could have been in for a long wait. Although I was relieved that it was well lit, it was only marginally warmer than outside. I couldn't summon the energy to open my suitcase to pull out more clothes. I was staring at my feet in despair, thinking I'd die of hypothermia, when a group of men entered and sat nearby. One of them, Callum Grieve, a climate change consultant who has since become a good friend, kept looking at me. Eventually he asked, "Are you going up to Arctic Basecamp?" When I nodded miserably that I was, Callum replied, "You must be Vanessa. We were searching for you, and we couldn't find you. We're so pleased you're here."

Callum lent me his coat and, after the funicular finally made its way up the mountain, he helped me carry my suitcase. He'd decided that I really needed to thaw out rather than spend a night in a tent in subzero temperatures. Some of the staffers from Arctic Basecamp gave me warm clothes and sat me by the fire in the hotel lobby. Finally, I was shown to my room, where I was advised to have a warm bath, drink some tea, and go to bed with a hot water bottle. I have never looked forward to bed like I did that night.

When I woke up in the morning, my hands no longer hurt, and all eight fingers and two thumbs were still attached. And the view of the Alps was spectacular.

That day was full of media interviews and spending time with the other activists also invited to join Arctic Basecamp: Wenying Zhu, who'd led the Environment Protection Volunteer Team at her university in Shanghai, China; Kaime Silvestre, a law student from Brazil, who was campaigning on behalf of the Amazon rainforest; Sascha Blidorf, who'd coordinated FFF (Fridays For Future) protests in Greenland and had run for a seat in the Danish Parliament; Brix Whiteman from England, who was volunteering at Arctic Basecamp with his brother; and Eva Jones from the US, who advocated for homeless women. I'd met Kaime before, at the UN Youth Summit in New York. Unlike me, he'd really come prepared for the cold.

Together we traveled down to the village of Davos, where the scientists from Arctic Basecamp spoke about how climate change was transforming the Arctic. Over the last three decades, the Arctic has warmed at a rate twice as fast as the rest of the planet,[1] leading some to predict that the Arctic will be ice-free by 2035.[2] Although our group was in Davos, we weren't attending the World Economic Forum itself (we hadn't been invited). Instead, we engaged in "outside" advocacy and awareness-raising. We visited a local school, where we met some of the students, who ranged in age from ten to fifteen. I was impressed. The kids were curious about my activism in Uganda and were eager to learn how they, too, could become climate activists.

That Friday, January 24, I was invited to speak at an FFF press conference with European climate activists Isabelle Axelsson and Greta Thunberg from Sweden, Loukina Tille from Switzerland, and Luisa Neubauer from Germany. We then participated in a climate march through the streets of the resort town, attended by hundreds of people who'd traveled to Davos to demand that WEF participants make the climate crisis their priority. Afterward, I joined some of the climate activists for lunch. That's when I saw the photo from the Associated Press (also known as "the AP") featuring Isabelle, Greta, Loukina, and Luisa . . . but not me. A photo had been taken of the five of us standing together. But when it was published by the AP, I— the only non-white activist—had been cropped out.

In the hours after the photo, I had little time to process how I felt. I'd been running on adrenaline and was bone-tired. I posted a tweet from the dining area asking the AP why I'd been cut out of the photo. But I needed to return quickly to Arctic Basecamp because most of the activists, including me, were leaving Davos that night for our home countries. The conversation there was muted; everyone was busy packing.

Arctic Basecamp's staff expressed surprise about my being cropped out and said they were sorry. But they didn't issue a statement or lodge a complaint with the AP, as I would have liked them to have done. That put the onus on me to say something, even though I was so upset that I didn't feel like talking about it.

From the Arctic Basecamp, I decided to record a video describing what I thought had happened. I was definitely emotional, and although I'd waited to do the live stream until I could compose myself, I couldn't refrain from crying. "You erasing our *voices* won't change anything," I said. "You erasing our *stories* won't change anything." And then I added something very personal. "I don't feel OK right now. The world is so cruel." As I tweeted at the AP on Friday evening: "You didn't just erase a photo. You erased a continent."

So much was happening it was difficult to grasp it all. A few hours later, I was on a train back to the Zurich airport, and then on an overnight flight home. Wi-Fi wasn't always available, and my phone battery was running down, so I couldn't follow the response to my tweets and video in real time. It was clear from what I *could* access, though, that my being cropped out had struck a chord, and many people around the world shared my outrage. I could also see retweets and posts from other climate activists and friends from home sending me good wishes and encouragement. The journey back to Uganda, however, gave me a chance to think more, and to figure out why the photo-cropping evoked such a strong reaction in me. As my flight neared Entebbe Airport, the words that came to me to describe how I felt were *heartbroken* and *depressed*.

*   *   *

Even now, well over a year after being cropped out of that photograph, it's hard for me to talk about what happened, and the hours and days that followed. I won't deny I felt personally humiliated, as though I was invisible, and that I'd wasted my time in Davos. I admit that being left out of the photo played into my doubts about my worth as an activist and whether I brought anything of value to the fight for climate awareness and climate justice. And I *was* stretched thin: in Davos, I was far away from home and engaged in a constant struggle to stay warm. In addition, I'd spent the previous few months attempting to absorb a dizzying array of new experiences, while keeping up my commitments to Fridays For Future strikes.

I know some people have wondered why what happened was such a big deal. After all, the same day the photo was published, then-AP executive editor Sally Buzbee issued a statement, recognizing that I as "the only person of color in the photo" had been cropped out, which she termed an "error in judgment."[3] She then sent an apology via her personal Twitter account. The AP replaced the photo with another where I *was* visible, and shared more photos they'd taken at the press conference, in which I was seated in the middle of the five of us.

The AP's director of photography insisted that cropping me out hadn't been a deliberate act of erasure, but was done on "composition grounds" by a photographer under intense time pressure: I'd been cut because the building behind me was distracting.[4] But, aside from the fact that the cropped photo still contained two other buildings, the question is, Distracting from *what* or *whom*? The Alps in the distance? My four white, European colleagues who were standing in front of the mountains? Or Greta herself?

The tears I'd shed in my Twitter video were not only ones of personal sadness, but of frustration and anger too. I was frustrated because the article accompanying the photo was titled "Thunberg Brushes off Mockery from US Finance Chief," and the AP had not only removed my photo but removed me from the list of participants.[5] It hadn't included a single one of my comments from the press conference where the five of us had spoken. This was deeply ironic, since, at that press conference, in addition to urging the delegates to break out of our comfortable addiction to fossil fuels, I'd asked journalists to reach beyond the comforts of their normal reporting: "It is time to report stories from every part of the world," I'd said, "because people are suffering from every corner of the world." Greta, too, had urged the journalists there to direct questions not only to her but also to the rest of us.

I was also frustrated because the AP generates two thousand stories a day and a million photos a year. It operates in 250 locations and reaches more than half the world's population—particularly in places with limited access to world news. By cutting me out of the photo they'd originally sent to global media organizations, the AP had denied an African activist a chance to be seen and, possibly, her message acknowledged.

I was angry because there were photos from other agencies and news outlets in which I *had* been included. That made me feel that the decision by the AP to leave me out was no accident. It wasn't the result of the photo being too large to download or the wrong size for newspapers. No, it was as if someone had determined I was the odd one out, an aberration, and that the photo wasn't satisfactory if I remained in it. Five of us were standing in a row; but there were two "ends" of that row and only one was shortened.

As the outcry on social media grew and the story of the photo-cropping made its way to the mainstream media, the AP was asked to respond. While they *did* express regret, I remain skeptical. I never received a formal message of apology, just a tweet. I also knew that the AP had released the uncropped photos only after the uproar.

* * *

Once I'd touched down at Entebbe, I was amazed to see I'd received more than a hundred messages from media outlets over several platforms wanting to ask me about what had happened. The days that followed my return home were filled with many interviews with both the international and Ugandan press. I was pleased that people from Uganda and other African countries, as well as activists from Europe and the United States, were tweeting at the AP, and my friends and many activists were reminding me to stay strong. My family, to be honest, couldn't understand why I was so upset, and tried to comfort me by playing down the incident. I don't blame them for that. I realize they wanted to protect me and cheer me up. This was difficult to do: as I explained to them and also discussed in media interviews, I felt like the photo-cropping was a direct expression of racism and sexism.

Here's why.

After the BBC reported how distressed I was,[6] some people on social media commented that I was being ridiculous, a crybaby, a snowflake, egotistical. They said that I should recognize how privileged I was. After all, I'd been invited to Switzerland and had spoken at a press conference, and yet I was complaining about being left out of a photograph. I shouldn't have stood at the end, some suggested; one troll wrote that I must have been cut out because I wasn't attractive enough. Some accused me of using the racism "card" to milk people's feelings and to get attention. Others argued that the reason I was excluded was because my work and opinions must not have had much merit.

But reacting with anger and emotion to racism isn't ridiculous. Injustices have to be called out. If you express your feelings, I see no reason why that lessens the reality or the truth of the injustice. In these comments, I also perceive an all-too-familiar gendered stereotype: women are irrational and overly emotional. Often, when we speak out, we're accused of being unreasonable in our expectations or our demands. When we say something forcefully, we're not considered commanding, like men, but shrill; we're not passionate, but nagging or bossy. In fact, to be considered a good African woman often means not saying anything at all. So, if we raise our voices, like I did, we're accused of not only being un-African, but of being unfeminine too. Some African commentators online were withering. I was told I needed to "know my level"—as a woman, as a Black woman, and as an African. "What did you expect?" wrote one. "You're a Black girl."

As for being privileged, it was *because* I recognized how fortunate I was to be invited to Arctic Basecamp (and New York and Madrid) that I was so distressed by the photo-cropping. I felt I was carrying with me a message on behalf of those Ugandans and other Africans who couldn't be at the UN Youth Climate Summit, COP 25, or in Davos. They were Africans who couldn't speak to the world's press or the powerful elites who'd gathered at these places and say that their lives mattered too.

I also wanted to speak on behalf of other FFF activists in Africa, like Davis Reuben Sekamwa, Nyombi Morris, Elizabeth Wathuti, Adenike Titilope Oladosu, and many others who couldn't be there. In erasing me, the AP erased climate activists across the continent who were trying to show that the climate crisis *was* an African issue; along with the fact that Africans were being most affected.

Why I was one of very few African climate activists in Davos is a good question. As Hilda Nakabuye said when she spoke at COP 25, "I do not understand why the most affected countries are always underrepresented . . . Voices from the Global South deserve to be heard."[7] Until many of us are in these spaces, most of us will continue to feel an obligation to speak out and be noticed—not for ourselves alone, but for the millions of people in our countries on the front lines of the climate crisis *right now*, who are suffering privation, dislocation, disruption, and death.

In addition, I'm not sure there's a single Black person, no matter where they are from, who hasn't walked into a room and once they've looked around and seen they're the only person there who looks like them, hasn't felt in some way they're representing every Black person on the planet. It's not a burden we're born to or we expect to carry when we are young; it's often a burden placed on us by others, who want us to conform to or counteract their prejudices. Sometimes we feel the weight of it and resent it. Sometimes it fills us with pride.

And sometimes it can leave us discouraged. I began to read comments on my social media posts that showed me that the photo-cropping had actually reinforced a dismaying narrative— almost the opposite of the one I'd been trying to advance. That narrative was that only white Europeans were worried about the climate crisis, because the climate crisis was the only thing white Europeans cared about. This was an extension of the wealth and privilege of a white world that had never really been concerned about Africa or the lives of Black people.

The comments from Africans and other people of color were very direct: "The climate change agenda is for white people," said one. "Leave to whites what is for whites," wrote another. "White people destroy the climate . . . and put other white people to fight climate change, all to make it look good for them," ran one tweet. This was, commented another person, "White people shit."

More insidiously, by cropping me out of the photo, the AP confirmed a suspicion some Africans held that whatever white Europeans might say, to the international media the climate crisis *was* a white European issue (and maybe one for white people everywhere). Some wrote I should forget about working with "white media" and only support Black-owned or Black-staffed news outlets: "This maltreatment of Africans will not stop as long as we Africans continue to allow the West to write our own history," wrote one. "Where was the African press?" Others told me that nothing would change and I should move on: "I as a Black man and African am used to this sort of exclusion," someone wrote. "I have learnt to move ahead and not care about it." One commentator despaired: "This is so sad . . . we've been cropped out all our lives. When will this stop???"

These comments essentially boiled down to a harsh reality: that to even ask questions such as these; to complain of exclusion or to demand inclusion; or to suggest that the media places white saviorism at the center of its lens, is to risk being thought of as ungrateful, impolite, a troublemaker, overly sensitive, prone to hysteria, or worse.

In his analysis of the photo, Dr. Robert Bullard, an African American professor known as the father of environmental justice in the US, told the *Guardian* that "climate activism among youth is perceived by the larger society as a 'white thing.' The un-cropped photo didn't fit the model," he said. "Racism has the intended purpose of making people of color invisible."[8]

The photo-cropping was classic "white saviorism," wrote doctoral student Chelsea McFadden in the *Journal of Sustainability Education* later in 2020: "The idea of the white savior is that there are people suffering in the world, which is codified as third world and racialized in the context of climate change, and that the only ones who can fix it are white people."[9] She continued, this explains "the rise in popularity of white climate change activists at the expense of activists like Nakate, who are ignored or even actively pushed out of the conversation."

This goes to the heart of why I think the photo was a big deal. By cropping me out, the AP had added to the mistaken belief that, as I said in one of those interviews, "African climate activists were absent from Davos; that Africans weren't active in the climate change movement; and that there wasn't a global youth climate movement that included people like me and many others in Africa, Asia, and Latin America."[10]

*   *   *

It's important for you to know that after the photo-cropping I received enormous support and solidarity from many climate activists, and do so to this day. Greta Thunberg called the crop "totally unacceptable in so many ways." Isabelle Axelsson added that my voice is "just as, if not more, valuable than ours in a place like this [Davos]."[11] Jamie Margolin observed that "racism, classism, and the erasure of marginalized voices isn't new." She added that "A photo crop-out is an easy way to describe it, but it's really a metaphorical crop-out from the narrative of climate science in general."

Jamie said that she had learned from what had happened to me: "[Vanessa's] experience made me reflect on the conferences where the picture didn't include the dark-skinned activists next to me—and I realize now I should have spoken up."[12] The AP itself also claimed that the incident had led to some "soul-searching." Sally Buzbee told a reporter for the AP: "I realize I need to make clear from the very top, from me, that diversity and inclusion needs to be one of our highest priorities."[13] The AP even announced that it would expand its diversity training as a result of the "terrible mistake" of the photo-crop.

I suppose I could, as that African man had recommended online, have grown "used to this sort of exclusion . . . learnt to move ahead and not care about it." I could have decided to say nothing, just continued my packing, left Davos, and carried on with my activism. But I *did* care, and I didn't see why Africans should settle for being excluded or ignored . . .

. . . Or interchanged. In their reporting after the cropping, both Reuters and the *Guardian* confused me with girls' rights activist Natasha Mwansa from Zambia, who'd attended the World Economic Forum, which I had not.[14] And erasure has happened to many more visible and more important people than me. At the G7 Summit in 2019, the Associated Press had posted a photo of Justin Trudeau (the Canadian prime minister), Narendra Modi (the prime minister of India), Emmanuel Macron (the president of France), and an "unidentified leader." Trudeau, Modi, and Macron were identified with their Twitter handles, but the "unidentified leader" was not. He was Cyril Ramaphosa, the president of South Africa, and his Twitter handle is @CyrilRamaphosa.[15]

If I hadn't spoken out, the AP wouldn't have released more of the photographers' images. The chance for a deeper discussion about race and the environmental movement might have been missed. And possibly fewer African voices, my own and others, would have been lifted up. Since the photo-cropping incident, I can see that journalists have been searching out activists on the African continent, and giving more exposure to the effects of the climate crisis throughout the Global South as a whole, including in communities on the front line if not the front page. This is certainly not all down to me. Far from it.

But I do think what happened to me, and my response to it, ignited a conversation that was long overdue.

Personally, the crop enabled me to push back against some of the reactions of my family, friends, and peer group, which extended beyond concern for my well-being to social respectability. It also allowed other family members to explore different narratives. If the photo-cropping had happened to her, my sister Clare told me, she wouldn't have spoken up. When I asked her why, she replied that she believed that even if she had, nothing would have happened. That's reason enough to make me glad I said something.

I'm not naive about how the world works. My specific erasure and my reaction to it increased my visibility on social media sevenfold and brought me to the attention of more of the local and international press. I spoke via several Ugandan outlets, and those journalists pushed the question of racism, which allowed me to talk about climate justice. That visibility led to further interviews, more invitations to speak, and ultimately to the book you're reading. I'm very aware that cynics might accuse me of shedding crocodile tears all the way to the bank. But as I've learned throughout my climate activism, to speak out is to risk a lot.

Being cropped out of that photo changed me. I became bolder and more direct in how I talk about the climate crisis and racism and how I articulate the many ways families are being impacted right now. It also changed how I thought about my career. Partly as a result of the pandemic, and partly as a result of my increasing commitment to climate activism, I avoided working at my family's shop until my father realized I wouldn't be returning. I was no longer interested in studying for a post-bachelor's degree certification in Marketing or an

MBA. I decided, from my perspective as a young African woman, that I would dedicate as much of my time as possible to addressing the many interlocking facets of the climate crisis, environmental justice, and gender discrimination—and to do so without apology or fear of erasure.

5

We Are All Africa

In October 2019, the Rotary Club of Bugolobi asked me to talk on the environment and climate change. My father, who was a member and had recommended me to the Bugolobi branch, often stopped by the hotel where the Club met after a day's work. So, I thought I'd surprise him and not tell him I'd accepted the invitation.

I looked forward to the opportunity. It would be the first time as an activist that I'd be addressing Ugandan professionals, many of whom were my parents' age. The audience would be civic-minded middle-class men and women who could raise awareness about the climate crisis and put pressure on the government and the private sector. Or they could do exactly the opposite: resist any change they perceived as slowing down what they considered "development" or "progress," and dismiss the concerns of the younger generation. My father would finally also be hearing me speak in public.

I needed to prepare. I watched videos of Greta Thunberg's speeches and remembered how she'd addressed the delegates at the Climate Action Summit in New York just a month before. I was struck by how direct and clear she was when she talked to people so much older than her and with significant power to bring about change. She possessed a kind of radical honesty that spoke from the heart, yet her speeches were all about science and facts and policy.

On October 11, almost an hour before my presentation was due to begin, I arrived at the hotel and took a seat at the back of the room, where I concentrated on what I wanted to say. After I was introduced to the club members, my presentation took about twenty minutes, followed by lots of questions from the audience. The Rotary Club presented me with a certificate of achievement, much like the ones my father had received, which made me smile. (As it turns out, my father couldn't make it to the Club that day, and so missed my speech. But he told me he received positive feedback about it from his friends. He finally got a chance to hear me speak when I gave a virtual presentation to the Rotary Club during the pandemic lockdown in spring of 2020.)

The questions that the thirty or so members of the audience raised at the meeting are common among many educated people in Uganda. Most expressed surprise at what I was saying. Everyone recognized there was something called climate change, but they'd never heard anyone take the time to present evidence of its existence, articulate its scope, or demonstrate how serious the crisis was. They seemed surprised to hear this information from someone so young who wasn't an expert, but were pleased I'd helped them understand the urgency of the problem.

At one point, a man said how puzzled he was that the ongoing degradation of the Amazon rainforest was widely condemned, even in Africa, but that no one was talking about the destruction of the Congo rainforest. As the meeting came to an end, that man's statement lingered in my mind. Why *weren't* Ugandans talking about what was happening in the Congo rainforest, especially since the Democratic Republic of Congo (DRC), in which about 60 percent of the rainforest lies, borders our country to the west? Since I had no good answer to that question, I decided I'd find out more about the rainforest and the dangers it faces.

What I learned shocked me. Like the Amazon rainforest, which spans national borders, the Congo rainforest or, more accurately, the Congo Basin Rainforest Ecosystem, stretches into parts of six countries: the DRC, Cameroon, Equatorial Guinea, Gabon, the Central African Republic, and the Republic of the Congo. The forest, known as the world's "second lung," is, like the Amazon, rich in biodiversity. It's also vital as a global carbon sink, sequestering up to 600 million metric tons more carbon per year than it emits—the same amount, says the World Economic Forum, as "one-third of the CO_2 emissions from all US transportation."[1]

The forest is also home for as many as 150 ethnic groups, including Indigenous Peoples such as the Batwa, Bambuti, and Ba'Aka. Humans have lived in the forest for more than 50,000 years and 75 million people today depend on it to survive. The ecosystem contains 10,000 species of tropical plants, several of which have been found to be useful in treating cancer, and many more may provide medicinal benefits. The forest also is home to a thousand species of birds, 700 species of fish, and 400 species of mammals, including gorillas, elephants, okapi (a cousin of the giraffe, but with a much shorter neck), and the black colobus, the most endangered species of monkey in the world.

Like the Amazon, the Congo Basin is being exploited for its

resources. Unfortunately, in the case of the Congo, this isn't a recent phenomenon. In the sixteenth and seventeenth centuries, the Kingdom of Kongo lost four million men and women to the Atlantic slave trade. Between 1885 and 1908, King Leopold II of Belgium ruthlessly and violently ravaged the region and its people for the production of rubber. The DRC itself became the staging ground for a proxy war between the West and the Soviet bloc at its independence in 1960. Almost four decades later, what was then called Zaire saw regional governments, including Uganda's, fighting one another for almost ten years after the fall of the US-backed dictator Mobutu Sese Seko in 1997. It's estimated that, through starvation or disease, this conflict took the lives of 5.4 million people. Even today, violence mars the Ituri, Kasai, and Kivu regions of the eastern DRC.

The ongoing political destabilization has certainly led to some environmental destruction, but 84 percent of the total deforestation is due to traditional slash-and-burn agricultural practices. Between 2000 and 2014, an area of forest greater than the size of Bangladesh was cleared in the Congo Basin.[2] Maddeningly, in 2020, deforestation rose globally by 12 percent, including in many countries in the Congo Basin region, even though most economies were in lockdown due to the Covid pandemic.[3] In the DRC, Cameroon, and the Central African Republic, forest loss was higher in 2020 than levels documented for 2019. How could this be? The data showed that too many countries were moving in exactly the wrong direction.[4]

Such losses are not just driven by local peoples. Industrial palm oil production, while still small, is increasing in the region. Chinese demand for wood, which is then turned into furniture for export (most of it to the United States), is also driving the cutting down of trees. The logging roads that companies build to transport timber open up previously inaccessible parts of the forest to large-scale hunting, poaching, and conversion of land to agriculture. These, in turn, degrade the ecosystems and cause further loss of trees and wildlife.

Logging also leads to speculation for precious metals and minerals. Columbite-tantalite (coltan), of which 80 percent of the global supply is found in the Congo, is a component of electronic circuit boards and computers, smartphones, and games consoles that we all use. In addition to causing pollution, mining of coltan usually means long hours, low pay, and harsh, hazardous conditions for workers, and mines have been linked to child labor and sexual exploitation of girls and women. Coltan, along with tin ore, tungsten, and gold, are so-called "conflict minerals," since gaining access to them is one of the major causes of fighting among militias in the region.

All of these factors, as well as population increase and long-term drying trends caused by climate change, have led scientists to calculate that, unless something shifts dramatically, the entire primary forest—all 500 million acres (202 million hectares) of it—may be gone by 2100.

The more I discovered what was happening to the Congo Basin, the more upset and angry I became. My first reaction was, *Why wasn't I aware of this?* Well, one reason is because the world's financial resources, including the media, are concentrated in the Global North. So, what's shown on television, published in print and online, and shared on social media is overwhelmingly oriented toward the developed world. So, for instance, during 2019 and 2020, we learned about the terrible fires in Australia and on the West Coast of the United States, even in Uganda.

As the Rotarian had observed, we're well informed about the deforestation in the Amazon and the many fires that have been lit to clear the land for grazing cattle, planting feed crops like soybeans for livestock, cutting timber, and mining. In some ways, we're more aware of biodiversity loss and how the Amazon's indigenous populations are being forced from their traditional areas than we are about the biodiversity loss and original inhabitants of the Congo. That's another reason why I was angry and upset. A fire in the Congo rainforest is as destructive as a fire in the Amazon, yet one was making news headlines, the other wasn't. If we couldn't defend the largest forest in Africa, I thought, then how would we protect the smaller forests, including those in Uganda?

A few days after the Bugolobi talk, I began my first strike for the Congo forest, urging others to join me with their placards (SAVE THE CONGO RAINFOREST), take a photo, and spread the message online about this vital ecosystem. My first results for this protest weren't encouraging. I discovered that not only had few people heard about the environmental and human tragedy continuing in the Congo, but some weren't aware the forest even existed. (At least I knew *that*!) It was a stark reminder of how someone might be able to crop an entire global ecosystem from the picture, and few would even know it had gone.

Finally, on Day 15 of this protest, when Greta retweeted a photo I'd taken of that day's strike, I began to receive some interest and momentum. More people began to share their photos, and on November 8, more than a thousand people joined in the strike. It was gratifying to see other FFF (Fridays For Future) activists taking part, including my Ugandan colleague Nyombi Morris, who'd joined me on the school strikes, and

Remy Zahiga, who's a geologist and indigenous rights activist from the eastern part of the DRC.

Remy is the founder of CongoEnviroVoice, an organization of young Congolese committed to the Basin's preservation, and advocating and agitating for the protection of its flora and fauna. Remy and I would address an online Greenpeace gathering in May 2020, in which he urged "leaders at the national and international level to turn their eyes to [the Congo forest] by supporting and respecting the agreements signed on the protection of wildlife and National Parks." He also called for such agreements to honor the rights of local peoples, while improving the security situation in protected reserves and parks.[5]

*   *   *

Of course, the destruction of the Congo rainforest was only one of the many interconnected disasters that climate change was exacerbating in Africa. In September 2018, Cape Town in South Africa was within ninety days of running out of water after three years of poor rainfall.[6] In March and April 2019, cyclones Idai and Kenneth struck the coast of Mozambique in the southeast of Africa, resulting in 2.2 million people needing urgent aid because of flooding; this in a country where 815,000 people were already in dire straits because of drought.[7] Malawi and Zimbabwe were also hit by the cyclones.

In August 2019, floods affected more than 200,000 people,[8] most especially at Niamey, the capital of Niger, as the Niger River rose by almost a meter, causing deaths and resulting in houses being inundated. Other countries in the region, including Nigeria, Central African Republic, Mauritania, and Morocco, also experienced major flooding around this time.[9]

In November, Djibouti, in the Horn of Africa, recorded two years of rainfall *in a single day* and several children were killed,[10] while in Kenya landslides and downpours took the lives of thirty-seven people in the West Pokot region bordering Uganda.[11] In May 2020, torrential rains washed away an entire town in Somalia, and killed hundreds of people there, as well as in Kenya, Rwanda, and Uganda, where a hospital in Kilembe in the west of the country, near Kasese town, was swamped by the ferocious currents. Pharmacies and a mortuary were swept away too.[12]

It wasn't only too much or too little water that overwhelmed the continent. In 2018, heavy rains during several rare cyclones, following a drought in the Arabian Peninsula, likely spawned a plague of locusts, the worst outbreak in seventy years.[13] The locusts spread to Ethiopia, Eritrea, and Somalia during the summer of 2019, and arrived in Kenya, Tanzania, and Uganda in February 2020, with another swarm threatening farmers in the northeast of my country in April.[14] The locusts destroyed 170,000 acres (69,000 hectares) of crops across East Africa in a twelve-month period,[15] putting millions of people who were already food-insecure at risk of famine.

If the years 2018 to 2020 weren't hard enough for the all-too-often forgotten regions in Africa (in a report, the international humanitarian agency CARE concluded that of the ten most underreported humanitarian crises in 2019, nine were in Africa), scientists are projecting that in the next several decades the extremes will become worse, as the global mean land temperature rises beyond its current 1.2°C (2.16°F) above pre-industrial levels. All but one of the past twenty years in Africa has been hotter than any year on record, and the new "normal" is hotter than any temperature in the recorded past.[16]

This is of particular significance since higher temperatures will lead to more evaporation and therefore cause more frequent and intense storms, potentially a wider spread of disease, and more drought. Continued warming of the Indian Ocean is forecast to generate more and fiercer cyclones.

Warming temperatures will also mean more food insecurity. A 2017 report by Future Climate for Africa, a South Africa–based climate science organization, projects a 10 percent reduction in the productivity of crops in sub-Saharan Africa, along with a drop of up to 50 percent in available water across large areas of southern and west Africa.[17] These present enormous challenges for countries like Uganda where a majority of the population relies on agriculture for its principal source of income. The Intergovernmental Panel on Climate Change (IPCC) estimates yield losses by 2050 of maize (corn) to be 22 percent across sub-Saharan Africa, with those in South Africa and Zimbabwe being more than 30 percent.[18] Of those crops that will survive the increasing temperatures, the authors of "Future Climate Projections in Africa: Where Are We Headed?" suggest that rising levels of CO_2 in the atmosphere will negatively affect their nutrient content, "resulting in serious protein and micro-nutrient cold spots in parts of sub-Saharan Africa."[19]

So, what would a 1.5°C (2.7°F) increase mean for the African continent? In blunt terms, it would be devastating. Researchers estimate that 1.5°C (2.7°F) would double (to six) the number of annual heatwaves in Africa by 2050,[20] and may result in 350 million people living in "mega-cities" around the world where the high heat could kill them. Lagos in Nigeria, Ivory Coast's Abidjan, and Khartoum in Sudan are among them. According to a study in the (US) Proceedings of the National Academy of Sciences, a 1.5°C (2.7°F) temperature rise will subject Lagos to a heat-stress burden *one thousand* times what it was in the recent past.[21] That would mean more demand for electricity, more need for water, and more deaths. And this in a country where half the population already has no access to clean water.

Kaossara Sani, a climate activist from Lomé, Togo, 186 miles (300 km) west of Lagos, is very aware of the human and environmental consequences of the climate crisis for her city and country. Kaossara had been volunteering to help homeless children when she encountered a nine-year-old boy from the countryside in the marketplace. He was living alone on the street, collecting plastic packaging to earn money, and wasn't in school. Kaossara tried to find an NGO or government agency to help him, but couldn't find one. She looked for him again, but he'd disappeared.

"I thought to myself, 'This young boy's life is destroyed: like that,'" she told me. She couldn't understand how or why parents were sending their children from their home villages to the cities to beg. Then she found out the answer:

> I realized that in rural areas, the main activity is agricultural. People depend on nature, and with climate variability and with floods, they can't support their family. They can't have good crops at the end. So, the only way they have is to send their own children to the city.

For Kaossara, speaking out about the climate crisis became a matter of advocating for children like this little boy. "Climate change is stealing their lives," she says, "not their future. It's already stealing their present."

Kaossara began the Act on Sahel campaign with other climate activists to help pay for seeds and fertilizer for farmers, who, as she says, "are on the front lines of climate change. Because if we don't do it, who will?" Act on Sahel also raises money for plant-based sanitary products and advocates for access to clean water and renewable energy. Like me, Kaossara speaks in schools about climate change, and encourages students to plant trees.

For the citizens of Lomé, where a quarter of Togo's population of eight million live, and the next largest city, Aneho, the issue of most concern is erosion along the 31-mile (50-km) Togolese coast. The seas are eating about 16–32 feet (5–10 metres) of shoreline annually.[22] "I'm living near the sea," Kaossara says, "and every day we are seeing how it's advancing." Togo's president, Kaossara told me, has questioned the meaning of "progress" or "sustainable development" if it means finding money to build walls to hold back the sea only to search for more money in a decade to build the walls higher or further inland.

Kaossara recognizes that her government is not responsible for fossil fuel–generated climate gases and is encouraging solar power–generated electricity. But she still sees failures and the importance of acting more broadly: "My government is not taking the climate issue seriously. The only way is to take action myself. If it is to plant maybe ten trees per month, it is better than nothing. If it is to maybe help only one person per month, it's better than saying that God will help us all. We can't change if we don't take action ourselves."

*　　*　　*

Kaossara is one of several West African climate activists focusing on the Sahel, the semi-arid region that stretches from Sudan to Senegal and acts as a buffer, preventing the expansion of the Sahara Desert into the heavily populated savannas to the south.

I got to know another West African, Nigerian activist Adenike Oladosu, in November 2019, when she, Elizabeth Wathuti of Kenya, and I were invited to a meeting in Ibadan, Nigeria, by the EET (Eleven-Eleven-Twelve) Foundation, an organization that promotes green solutions and job opportunities to encourage economic growth in that nation. It was my first time in another African country.

In Ibadan, Adenike told me about her campaign to draw attention to another vital African ecosystem: the Lake Chad Basin, which encompasses parts of Algeria, Cameroon, Niger, Nigeria, Central African Republic, Libya, and Chad. Since the 1960s, the Basin has shrunk from being the world's sixth largest inland body of water to less than a tenth of its original size.[23] This is the result of poorly managed irrigation, increased use by a growing population in the Sahel region, and the effects of climate change. Now the desert advances each year, and the Basin, which used to provide water and food for between 20 and 30 million people, is home to nearly 11 million people who require humanitarian relief.[24]

The shrinking of Lake Chad has also reduced agricultural livelihoods in the region, leading to extensive outmigration and more cross-border and regional conflicts. As Kaossara observes, the shrinking of access to fertile land and water lies at the heart of many local and regional conflicts in Africa, including between farmers and herders. "They were friends or family before," she says. "Now they are killing each other for these resources, and some, who don't want to, join terrorist organizations, or are still dreaming of going to Europe. Some will lose their lives on the open sea and some will try to find another place in other African countries."

Adenike is campaigning to increase awareness about the social, political, economic, and ecological crises caused by the drying out of the Lake Chad Basin. She considers it a wake-up call to the entire world about what happens when an ecosystem can no longer support the numbers of people who depend on it. She writes:

> A combination of decreasing rainfall, increasing temperature, and other climatic elements will destroy the economic livelihood of people, be they in Africa, Europe, or Asia. Lake Chad represents what the world will witness in decades . . . [That combination] will lead to the creation of internally displaced persons and camps, desert expansion, resource control, armed conflict, and, finally, failing democracies.

In some way, therefore, we are all Africa.

During those few days at Ibadan, Adenike, Elizabeth, and I fell into an easy conversation on how we could collaborate. Elizabeth told us about her project planting fruit trees in schools (which I discuss in Chapter 6), and Adenike described her work with young and older women in communities threatened by natural disasters, and the dangers of sexual violence and abandonment that they face as a result.

The three of us each faced similar difficulties. We recognized that many African voices were struggling to be heard—not only internationally but within the continent and even within our own countries. We were frustrated by how few ordinary people were aware that the climate crisis was behind so many of the disasters that they called "God's will," and how difficult it was to create a uniform message on climate action that would carry weight—in our countries, regions, and even globally.

Some problems lay beyond our immediate capabilities to fix, but we agreed on a few actions we could take together. We'd amplify one another's voices by sharing our work online, and emphasize to the international media the importance of the collective efforts of the growing number of climate activists we were in contact with. This way we'd show there weren't only a handful of people in Africa fighting for climate justice, and that we echoed the concerns of people, young and old, in many countries throughout the continent.

We led a climate strike at the University of Ibadan for Lake Chad and the Congo rainforest. Later, at my presentation at the EET event, during which Elizabeth was honored, I told attendees, "If no one is going to fight for Africa, it is because Africans are silent."

*   *   *

I had an opportunity to take my Congo strike to COP 25 in Madrid. I walked through the expo in which national governments set up pavilions to showcase what they're doing to promote a more climate-compatible future. You won't be surprised to learn there's a lot of glossy greenwashing at these expos, and COP 25 was no exception. After some activists and I searched in vain for the Ugandan pavilion, we came across the one for the Republic of Congo (Brazzaville). There, I talked to the people staffing it about the strikes I was staging on behalf of the Congo rainforest.

They were not pleased. They took it in turns to tell me that, since I'd never been to their country or seen the Congo rainforest, I had no comprehension of the needs of its citizens or the importance of developing the region. The people of the Congo required properly constructed houses, one of the men said, which I took to mean that the wood to build them had to come from the forest. It was a strange and unsatisfying discussion. Later, we held a strike for the Congo forests in front of the pavilion. As you can imagine, the staff I'd met earlier glared at us from a distance, and I'm sure they were happy when we returned to the conference center.

It's true, I haven't been to the Congo region, and I may not fully understand the developmental needs of the people who live in or around the Basin. But surely it doesn't make sense to destroy the world's "second lung" for furniture, palm oil, building materials, minerals, or fossil fuels.

Some of you reading this may feel it's presumptuous for any of us—whether Adenike, Elizabeth, or me, or any of the other African climate activists—to claim to speak for the whole of Africa, a continent made up of 54 states, home to 1.275 billion people, and encompassing hugely varied ecosystems, peoples, cultures, and social conditions. And I agree that it's absurd that one individual should presume to be, or even be considered as, the spokesperson for a continent. Yet, after the AP's decision to crop me out of the photo at Davos, almost every interviewer since has asked me not only how climate change is affecting Uganda, but what its consequences are for other parts of Africa. I'm aware that I can provide only a snapshot of what the continent is undergoing, based on what I've learned from other activists. And I recognize there are limits to what I can directly do to influence policy for the Congo Basin—or anywhere else, for that matter.

But I believe that we need to speak out—to "break the silence," as Kaossara says. I see my role in climate activism as bringing up conversations that many people have never had, and highlighting the destructive policies and investments of banks, hedge funds, multinational corporations, and governments—all of which would like the rest of us to have no idea what they're up to. I see my task as drawing attention to communities that people may not have heard of, where lives are being upended and lost on a daily basis.

No country, no matter where, is just a country. What happens in the Congo Basin rainforest doesn't just affect people in countries in central Africa; it influences weather patterns across the world. The climate crisis respects no geopolitical borders, political bloc, or regional trade associations. So, what happens in the Congo isn't just the business of the Congolese, or their neighbors. It concerns all of us.

Finally, I'll be the first to agree that we need more diversity on platforms and more young activists to be given opportunities to talk about the challenges their countries or regions are facing. There should be 54 or 216 or 1,028 activists from every African nation state speaking at international climate conferences and to their own governments. Every activist has a story to tell; every story has a solution to give; and every solution has a life to change.

6

A Greener Uganda

In October 2019, my colleague Hilda Nakabuye from Fridays For Future Uganda spoke at a meeting of C40, an international network of mayors of cities prioritizing climate resilience and climate action. Hilda had missed three months of school after floods engulfed her family's farm and her father hadn't been able to secure the funds for her tuition.

"After the massive effects of climate change in my home village," she told the mayors and their staffs, "the heavy rains, the strong winds that washed away our crops, leaving the land bare, the constant dry spells that left the streams dry—my parents had to sell off our land and livestock, to sustain our lives."

Hilda fought back tears before she continued: "I am lucky that I am still surviving, and I will not take this chance for granted, because people are dying every day." She explained how she'd become involved with FFF (Fridays For Future). "I made a decision to protect the only place I call Earth," she said, "and by this I join young activists from all over the globe to protect our future through endless fights [and] sacrifices."[1]

Hilda's ordeal, and that of her family, is more and more common in today's Uganda as the effects of global heating intensify.

Uganda's temperature has increased by 0.2°C (0.36°F) in each decade since the 1960s.[2] A 2016 report from Future Climate for Africa concludes that temperatures in Uganda will likely rise by 1.5°C (2.7°F) above pre-industrial levels as soon as 2030, and potentially by a catastrophic 3.3°C (5.94°F) by the 2060s.[3]

This is an emergency for my country since, although urbanization is ongoing and fast-paced, almost three-quarters of Ugandans still live in rural areas and, like Hilda's father, rely on agriculture as their main source of income. Nor is Hilda exaggerating about choices being a matter of life and death. For, in addition to forecasting more frequent and severe floods, scientists have calculated that rainfall may *decrease* by 7.5 inches (188 mm) by 2080, as the seasonal rains, upon which farmers plan when they sow and harvest, shift.[4] This will make it harder for the country to feed itself, and Uganda's population is projected to more than double, from 40 million today to 100 million by 2050.

The floods also carry public health implications. In Uganda, relentless downpours in October 2019 around Lake Victoria caused water levels to surge by May 2020 to a record 44 feet (13.42 meters).[5] Some of the more vulnerable, low-lying settlements around the lake, which also borders Kenya and Tanzania, were flooded: 200,000 people were displaced and the communities' drinking water was polluted, increasing the risk of outbreaks of water-related diseases like malaria, schistosomiasis, cholera, and dysentery.

It's not only informal settlements around the lake that experience flooding. On the Easter Monday holiday a few years ago, I'd gone with my siblings for a family getaway to what's called Miami Beach, where we planned to spend the day listening to music and swimming. When we arrived, though, the entire beach was flooded. We kept trying to find a dry place, but couldn't. I noticed the water was really dirty, most likely from soil or sewage that had been swept into the lake's water along with the rain. Soon after we arrived, we decided to return home.

Revocatus Twinomuhangi, a geographer at Makerere University, was asked in 2020 to comment on the consequences of rising waters in Lake Victoria and the climate crisis overall. "Climate change is becoming [more of] a reality than ever before," he said, "and if we do not stop human activities along the lakeshores, water catchment areas, and reserve forests, we are heading to some disasters that [African countries] have no resources to handle."[6]

Kampala's streets often flood during the rainy seasons. Many roads have poor drainage systems; many aren't paved; most have potholes. But the increased intensity of the floods and the many pools of deep water in the city have begun to worry me. One morning about two years ago, I was taking the two-year-old daughter of a friend of my mom's to church. We were late, so I decided we should catch a *boda boda* (motorcycle taxi). It had rained heavily the night before and soon we were on a road that had so much water that I couldn't see how we could pass. The driver assured us we could make it, but I couldn't risk the life of the young child. So I turned back, took a taxi, and we got to church that way—late, of course, but at least we were safe. I learned later that a young woman riding a *boda boda* that same morning on a street not far from where we'd been had fallen into the water. Some people rescued her and she was OK, but it was scary to know how easily that water could engulf you.

It's all too easy for people in Kampala to step into the road expecting to be on firm dry ground, but instead find they're actually in a ditch or pothole filled with water. Some of the ditches alongside the road and the potholes can be several feet across and sometimes sinkholes open up in the roadway. These can be large enough for a vehicle to fall into. I've seen stories on the news of people being sucked into the water and drowning. Now, whenever there's a heavy downpour and Kampala's streets flood and my mom or siblings or friends are in town, I worry. I'll call to remind them to be careful and to look where they're going. "Don't stand or step anywhere unless you're sure you can see land," I'll say. "It's dangerous. Please be careful." Even onlookers who might want to help rescue someone often can't help. It happens too quickly. It takes you by surprise.

*   *   *

Outside of cities, deforestation is a major cause of flooding and drought in Uganda (as it is elsewhere), and sadly it's just as common. Forests help regulate rainfall and stabilize local climates, but in addition when trees are felled in more mountainous regions, those areas are prone to landslides. Forests and trees help keep soil and air temperatures stable, maintain soil fertility by preventing it from disappearing as silt into rivers, and enhance the land's ability to hold water in dry seasons. With fewer forests and trees, weather disruptions become more frequent, and less carbon is stored in trees and more is released from the bare soil.

Over the last twenty-five years, nearly 7.5 million acres (3 million hectares) of forest in Uganda has been lost, according to the Uganda Forestry Authority,[7] which is an alarming reduction of 63 percent.[8] According to Global Forest Watch, between 2002 and 2020, 168,000 acres (nearly 68,000 hectares) of primary forest in Uganda was cleared. That's the equivalent of 14.3 megatons of CO_2 emissions.[9]

We tend to think of deforestation as a mass event, but even the loss of a single tree can have consequences. My parents own a small piece of land in my father's home village of Butega in Mityana District, 43 miles (70 km) west of Kampala, where we grow cassava, maize (corn), sweet potatoes, and types of bananas (members of our extended family keep an eye on the farm on a daily basis). One day, the next-door neighbor cut down a large tree on his property. As my father tells it, ever since, when the heavy rains come, the land floods and sometimes the crops are washed away because the tree roots can't hold the soil and absorb the rainwater. Not only does the next-door neighbor now have the same problem with his crops, but so have many others in the village—and all because one tree with deep roots was felled. "People don't know the importance of these things until they actually lose them," my dad said to my mom and me as we watched a news item one evening about climate change. "Only then do they see the damage they are causing."

Most trees are cut down in Uganda to produce charcoal, a common source of energy for cooking and heating. It's estimated that each year in my country, 6 million metric tons of wood is processed into 1.8 million metric tons of charcoal.[10] Although charcoal kilns are more efficient than they used to be, producing one metric ton of charcoal still results in emissions of three metric tons of CO_2 equivalent.[11]

Clearing forests also endangers wildlife and can lead to conflict between animals and people. Uganda still has several thousand chimpanzees living in its forests, although in some areas, their habitats are shrinking, and quickly. Chimps near Lake Albert in western Uganda, starved of the fruits they'd normally eat, as their forest habitat dwindles and agriculture expands, have been drawn to forage in local people's fields. Some have attacked and even killed young children. It's a tragedy for everyone. As biodiverse habitat is destroyed or encroached upon by human settlements, wild animals come into contact with humans—raising the potential for transmission of disease, as with Ebola or HIV/AIDS.

Even good deeds may have unintended consequences. Uganda has a long tradition of welcoming refugees and providing them with land, which is sometimes forested. In recent years, Uganda has taken in 1.3 million people, mostly from South Sudan, the DRC, Sudan, Somalia, and Burundi, countries which have been destabilized after decades of internal conflict. Uganda has rightly been praised for this hospitality. But, the UN Food and Agriculture Organization says, the refugees are adding to existing pressures on Uganda's forests and woodlands for fuel wood, charcoal, and timber. Unless efforts are made to reduce these demands, all of the forests could be gone by the end of the century, analysts say.[12]

Unfortunately, from what I have seen, the Ugandan Government is either actively supporting deforestation or neglecting to act quickly enough to stop it. In 2020, despite strong opposition from environmental NGOs, the Government's own environment authority gave permission for two sugarcane companies to carve thousands of hectares out of Bugoma Forest, which spans 100,000 acres (40,000 hectares) of the border between Uganda and the DRC. As part of #SaveBugomaForest, Leah Namugerwa has participated in campaigns to protect Bugoma, both online and in the reserve itself, and in September 2020, over twenty climate activists were arrested for protesting there.

Leah says, "Deforestation is the main root of all the climate crisis we're facing in Uganda."

In December 2020, the Government finally stepped in to stop the clearing of Mabira Forest in the east, after individuals and companies who'd acquired disputed titles to parcels of land in the forest caused the degradation of considerable sections of it. I've driven past Mabira Forest myself, and where once the stands of trees were thick, the interior of the forest has been hollowed out for agriculture.

For my friend and climate activist Evelyn Acham, the destruction of Mabira Forest is a symptom of what she calls "the biggest obstacle" to addressing the climate crisis in Uganda: a failure to work collectively at all levels of society for the common good. "Our system has no accountability. Every citizen needs to be heard; every voice matters; every sector. The leaders are looking at making money and people are not thinking to develop communally. If people's mindset could change, and people start thinking about their fellow citizens and their well-being, we could have a very big impact."

She continues: "If our leaders had a heart for all their citizens and love for our country, I don't think that Mabira Forest would be sold. Because that's the biggest and only natural forest we have in our country. But a big percentage is being sold off. Look at Bugoma Forest. It's currently threatened; they want to finish it off."

*   *   *

My family is fortunate not to have to rely on our farm for survival, although the food we grow there always tastes more delicious than what we can buy in Kampala. I also get a good feeling knowing where the food we're eating comes from, and it makes me even more aware of not wasting it. I also like contributing to getting the food to our table, although I'm glad I don't have to be a full-time farmer. As a city kid, I can tell you that planting and harvesting are hard work. I recall picking maize (corn) with my mom, one of her friends, and my brother Trevor. Because the plants weren't all together, we had to go back and forth pulling off the ears one by one and putting them in a basket. Pretty soon, Trevor and I were exhausted, and as it got later and later, both of us kept asking, "When are we going home?" I looked at my mom and her friend, and I could see they weren't tired, even though they'd been working for hours.

Tens of millions of Ugandans grow food this way, and our country actually produces enough to export some crops. Poverty nonetheless constrains people's access to enough nutritious food. The World Bank defines 20 percent of our people as living in extreme poverty, and more than one-third survive on less than US$1.90 each day.[13] Almost 30 percent of kids in my country younger than five years old are stunted, an indication that they don't eat enough to grow and develop fully, according to the 2019 Global Hunger Index,[14] and food insecurity is an ongoing problem. In November 2020, the World Food Programme (WFP) assisted more than 1.2 million Ugandans, both adults and children, with food rations or cash to buy it.

Perhaps not surprisingly, food insecurity is worse in rural parts of the country. Most farmers rely on income from cash crops and use almost all their land to plant them, foregoing the cultivation of leafy greens, varied grains, and fruits their grandparents might have eaten. The WFP is liaising with a local NGO to improve farmers' post-harvest practices and the marketing of beans, which are rich in iron. In cities, fruit and vegetable consumption is low, and it can be challenging to find affordable, ready-made healthy meals or to buy the varied ingredients to make them.

Current food insecurity will be made much worse as climate change intensifies. Uganda's population is urbanizing, and demand for animal-based foods is growing (as it does in most places, alongside increases in production). In a 2014 report, the FAO (Food and Agriculture Organization of the UN) found that even under best management practices, raising more animals for food and setting aside land to provide feed crops for them would lead to more depletion of Uganda's natural resources. That's in addition to the pressures of longer droughts, rising temperatures, desertification, and flooding. "In all scenarios," the FAO report says, "the competition for land, feed and water is fierce."[15]

A Uganda devoid of forests to regulate temperatures, soak up rainfall, hold topsoil, prevent the silting of rivers, and provide biodiversity; a country with average temperatures of 3.3°C (5.94°F) above pre-industrial levels, and with an irregular food supply and subject to increasingly catastrophic weather events—this is a future no child should have to face. But in Uganda, millions are.

*   *   *

To handle such multilayered and interconnected situations, I find myself looking for solutions that are actionable, scalable, and holistic. One such solution arose through what's now called the Vash Green Schools Project, which launched in October 2019. (Vash is a nickname my friends have given me.)

In August 2019, I was approached by Tim Reutemann, a

climate finance expert from Switzerland, who'd been liaising with the Swiss, German, and Ugandan Governments to provide solar panels and clean cooking stoves to schools in Uganda. That project had foundered, but Tim remained personally committed to it and was looking to channel private money to make it happen. It was important to him that the project be done in a way that was "real, verified, and additional", as required by the Paris climate agreement. Tim asked if I'd be open to collaborating. As a first step, he suggested I find five suitable local schools and work with the head teachers, community leaders, and vendors of solar panels and cooking stoves to develop a plan. I agreed. I immediately saw the value of this project. Many rural schools were without electricity. Green energy would offer them more options for studying, along with hands-on learning. Cleaner stoves for school meals would reduce demand for charcoal and save forests. It would save schools money too.

When I was a student, I'd had experience of what can happen when basic resources are scarce. During my last two years of secondary education at a girls' boarding school, we had water shortages a few times during the term. When this happened, every student would have to take a plastic jerrycan in one hand and a bucket in the other and collect water from a natural reservoir that the school had access to, about twenty minutes' walk away.

It wasn't a matter of choice: unless you retrieved the water, you wouldn't be able to bathe or wash your clothes over the next two or three days, until the water supply was restored. Even though we always had water to drink in the school's dining halls, the preciousness of that resource and the fact that the jerrycans and buckets held five gallons (twenty liters) of water each (that's a combined weight of 22 pounds/10 kg, not including the jerrycan or bucket) encouraged us to take conservation seriously. I recall walking back to my dormitory very slowly, trying to manage the weight and not let any water slosh out.

The electricity supply at the school was also subject to interruption—usually during the night, after which the generator would kick in. But if the power went out in the early evening and the generator failed to work, we'd be excused from our homework, which, to be honest, we weren't too unhappy about. By moonlight or using a flashlight (if we remembered where we'd left it the last time), we'd return to our dormitories. Of course, these were relatively trivial inconveniences in comparison to what some students have to endure. Many schools in Uganda still lack proper toilets or water for washing hands. This lack of sanitation increases the risk of diseases spreading, including potentially fatal bacterial infections or viruses.

When it came to solar panels, I also had some experience. Our family placed solar panels on the house we lived in when I was younger. Back then, the electricity supply was less reliable than it is now and we'd be plunged into periodic blackouts, sometimes as often as three times in a week. My family's shop also sold solar batteries.

I knew that solar offered many benefits. Maintenance costs were low and the remote locations of some communities, well off the electrical grid, made the panels a scalable and practical way to address energy poverty. Although Uganda has a lot of potential for solar-powered energy, few homes in Kampala have solar panels, and most purchases take place before Christmas, so people will have light in their villages for the festive season.

More efficient cooking stoves were also a promising solution. According to the Clean Cooking Alliance, burning firewood or charcoal is the primary source of energy for cooking and heating for almost half the world's population. This is still the reality for 90 percent of Ugandans.[16] Many schools have to buy firewood or charcoal to cook food, and kids in primary (elementary) and secondary (high) schools may be asked to bring wood with them for the school's use. Since not all firewood is gathered from plantations grown for this purpose, local trees are cut down, or branches are severed, which can permanently harm or even kill the trees.

In addition to contributing to greenhouse gas (GHG) emissions, burning firewood or charcoal causes nearly four million premature deaths worldwide *every year* from "childhood pneumonia, emphysema, cataracts, lung cancer, bronchitis, cardiovascular disease, and low birth weight," because of smoke inhalation. Twenty thousand of those deaths occur in Uganda; 5,700 of them are children.[17]

Here, too, I had some personal experience. When I was young or home from school holidays, I'd help my mother by lighting the charcoal in the cooking room that was attached to our house. The entire space would fill with acrid smoke for at least fifteen minutes. Twice a day, morning and night, the smoke would make my eyes burn and water, as if I were chopping onions, except worse.

By contrast, clean cooking stoves use less wood, are more efficient, and are less toxic. They lessen fuel consumption by up to 50 percent, which in the case of the stoves we supply to schools reduces the amount of firewood used per term from five truckloads to two. This reduces greenhouse gas emissions per school by an average of fifty tons of carbon dioxide per year.[18]

Until I heard from Tim, I'd thought of myself as an activist who spoke about climate realities in schoolrooms and not as an entrepreneur who'd act on climate change with the school itself. Even though I was a Business major, I wasn't a businessperson, and it never occurred to me that someone so far away would do such a thing. When I told my father about Tim's request, he was dubious, especially as Tim had said he'd not only fund the panels and cooking stoves themselves but cover my transportation and other costs. My father demurred: "Let's wait and see if he sends you the money," he said. "That would confirm he's serious." When the money came through to the new bank account I'd opened (my first), I was astonished and my father was also impressed, since, he said, it was very rare for white people to entrust Africans with their money.

As I considered which schools were good candidates for the solar panel and cooking stove installations, I opted to look at schools in Mityana. I took some videos of the schools and sent them to Tim, so he could see they'd benefit from a clean source of regular energy and proper cooking facilities.

My father recommended a bulb and panel supplier in Mityana he considered reliable and who could provide everything we'd need and manage the installation too. I've worked with him ever since. For the cooking stoves, I consulted my mother, who mentioned a neighbor who'd owned one when we lived in Luzira, a suburb of Kampala. It turned out her stove was the exact kind I'd thought of for the schools, and we now use the same supplier she used.

Our aim was to keep the budget for each school to under US$3,000, which paid for one large cooking stove (known as a "three-in-one"), a 200-watt solar panel, about twenty energy-efficient light bulbs, and a battery that could store 150 watts of power. Once Tim and I agreed on what we'd provide to each school, I had to sort out the installation. The process takes about a day for the panels, and three days for the cooking stoves, which have to be constructed on-site. Usually, we begin the installation on Friday and finish it by Sunday. If we do two schools at a time, they'll be completed by the next Monday. In choosing which schools to work with, we try to find those most in need.

To date, the Vash Green Schools Project has supported the installation of solar panels and the building of clean cooking stoves at a dozen primary schools in the Mityana and Wakiso Districts of central Uganda. I oversee each installation, usually traveling to the school with one of my parents, or with some climate activist friends. I try to stay in the background and am present only to make sure that everything's proceeding as planned.

Observing the process and its completion is gratifying. You can see how much it means to the students, teachers, and the community, as the old cooking area is demolished and a beautiful new one takes its place, and as the solar panels are fitted onto the school's roof. I remember vividly our first installation, at Marking Progress School in Mityana, in the autumn of 2019. On the day the installation was finished, students, parents, and teachers surrounded me. They were obviously delighted they'd have reliable electricity to aid study.

This was not the only benefit. The children would eat regular cooked meals, enabling them to concentrate more in their lessons. They'd no longer have to gather firewood to bring to school the next day to stoke the oven. Not needing to purchase so much firewood would mean the school could spend money on other supplies, fewer trees would have to be cut down, and GHGs would be reduced. The school would be a more secure and safer place because of the extra light. And the school chef would be preparing food in a less smoky environment.

Speeding and scaling up clean sources of power in schools and communities will reduce energy poverty and also contribute to Uganda meeting its Paris climate agreement targets. The Vash Green Schools Project and others like it also have the potential to directly benefit the lives of many girls and women in rural communities. The installations also give me an excellent opportunity to talk to the kids and the community about the climate crisis and clean energy. Teachers have told us they're aware of the risks of cutting down trees and the climate disasters that will follow, but without alternatives for cooking school meals, they worry the children will stop coming.

After one installation, the head teacher led us to the back of the school to show us the trees he'd planted, and told us they would have been used for firewood, but now he could let them remain. At another, we experienced the effects of recurring flooding firsthand. The road leading to the school had turned to mud, and the vehicle carrying our supplies and us became stuck. As a result, we walked to our destination, each of us in our small group carrying the panels, batteries, and cooking stove materials. The installation could continue as planned, which was the most important thing.

While I appreciate the thanks of the students, school staff, and community who may drop by during the installation, sometimes expectations are raised I can't fulfill. The installation in October 2020 at St. Mark's Primary School in Wakiso District coincided with the end of Sunday services at the nearby church. Several parishioners approached me and said varying versions of, "May God continue blessing you, so you can do more."

Others have come to me after an installation and said, "Thank you for this. But we also have *this* challenge." Schoolteachers and principals have asked me whether I can also supply computers, printers, and water pumps. All I can say is, "I'm not in the place to provide that right now. But if I receive the support, I'll definitely keep your school in mind."

Of course, I'm aware that every technology comes with environmental costs. It takes minerals, including rare earth ones, to make photovoltaic cells for panels, and then energy to manufacture and ship them around the world. Battery production and especially storage and disposal not only require resources but generate toxins. My father's business handles the latter directly: customers bring dead or spoiled batteries to his shops and they're sent to Uganda Batteries Limited for recycling.

It's my understanding that it's possible to use these old batteries to make new ones, maybe even for solar panels. But right now, every energy strategy is a weighing of alternatives over which impact is the least damaging in terms of GHG emissions—and it's obvious that solar and wind are much better options than continuing to use fossil fuels for electricity, transportation, and industry.

* * *

One additional component to the Green Schools initiative that I've encouraged is the planting of fruit trees on school grounds. I was inspired to do this by Elizabeth Wathuti of Kenya, who started the Green Generation Initiative to encourage kids in schools, along with their families and communities, to do just that. Elizabeth's objective is to foster a tree-growing culture, supplement kids' diets, and provide environmental education and stewardship. She was inspired to start this work by the late Wangari Maathai, who founded the Green Belt Movement, an initiative that has planted more than 50 million trees throughout Kenya and has encouraged similar efforts around the world.

Professor Maathai, who was awarded the 2004 Nobel Peace Prize, recognized the multifaceted value of trees: as conservers of topsoil and streams, as sources of fuel and fencing, as providers of shade and habitat for wildlife, as sequesters of carbon, as reducers of food insecurity and suppliers of essential nutrients, and—through the development of seedling-growing and planting cooperatives—as a means to empower women.

Elizabeth is from Nyeri in Kenya's Central Highlands, the same region where Professor Maathai grew up and that she later represented in Parliament. She now works as director of campaigns for the Wangari Maathai Foundation. Here's how she described the roots of her activism to me: "I realized things were making me angry, such as going to a place and finding whole stands of trees cleared and cut down; or going to a stream and finding it not as clean as the stream I knew from back at home. I developed a hunger to want to do something about these challenges, because I know at times people stop at the anger." That hunger led her to join the climate movement, "to use my voice to speak up against these things that are devastating for me" and "to find out how I can be part of the solution."

Elizabeth's Green Generation Initiative has trained more than 20,000 schoolchildren and planted over 30,000 trees in school compounds in Kenya. What's also amazing is that the Green Generation–trained kids have kept almost all of those trees alive. When children are excited about the survival of their trees, they're more open to talking about conservation, and they enhance their educational experience by beautifying their school surroundings.

Uganda has many local and international NGOs helping to plant trees. My father's Rotary Club has an Environmental Sustainability Rotary Action Group (ESRAG). Before the pandemic, ESRAG's Mission Green committed to planting five million trees each year in Uganda and Tanzania between 2017 and 2022. The project aims to engage nine million pupils in primary and secondary schools to plant one tree each.

Joining in and supporting tree-planting projects has become a popular way for politicians in Uganda (and in many other countries) to burnish their "green" credentials. That said, while planting trees indigenous to their regions is valuable, it's not a replacement for protecting existing forests or restoring degraded woodlands. It's also not a substitute for conserving grasslands (which also sequester large amounts of carbon) and ensuring that old-growth forests and woodlands remain intact.

In tandem with solar panels, clean cooking stoves, and reforestation and afforestation programs, Uganda needs to find other ways to lessen charcoal use and the burning of fuel for food. Uganda's nationally determined contribution (NDC) under the Paris Agreement is to reduce its GHG emissions by 22 percent by 2030. In his new role as a division mayor for Nakawa, my father is piloting the use of a machine that turns organic waste into briquettes that can be used instead of charcoal. These machines not only lower the amount of methane (a much more potent GHG than CO_2) that is released from landfills, but they eat into the estimated 1,300 tons of waste that Kampala generates each day. This is not a new technology, but its cost has generally been out of reach for ordinary Kampalans. That's especially true for the 87 percent of the citizens who live in informal settlements, the vast majority of whom also don't have access to electricity.[19] Producing these briquettes could provide Kampalans with a viable business.

I hope the Vash Green Schools Project will expand; Tim and I have set up a GoFundMe page to make it easier for people to contribute. However, whether nine or ninety, the number of schools that I can personally oversee is a tiny fraction of the 24,000 primary and secondary schools in Uganda that would benefit from solar technology and clean cooking stoves. At a minimum cost of US$72 million, such a goal remains far beyond my financial means, let alone my logistical capabilities.

Likewise, there are limits on what one person, or organization, or civil society, or even the private sector can do to meet the energy demands of an entire country. This is why the role of government, and governance generally, is so important. Though here, too, most governments are not doing what they should.

*   *   *

In 2006, Uganda discovered oil in the Lake Albert Basin, with potential reserves of more than six billion barrels as well as 500 billion cubic feet of gas, making it one of the biggest reserves in sub-Saharan Africa.[20] I was in primary school when the news was announced, and I remember our teachers being so proud and happy that Uganda had found "black gold." The find even became a common set of questions in school exams: *When was oil discovered in Uganda? What does it mean?* We all knew the answer to that second one: a fast track to success, jobs, economic development, and the reduction of poverty and unemployment.

A similar story with similar outsized expectations is being told today. In 2017, the Ugandan Government approved a multinational consortium's bid to build a refinery in Kabale, in the west of the country near the border with Rwanda, to produce kerosene, gasoline, heavy fuel oils, diesel, and other products.[21] It's estimated the refinery will yield 60,000 barrels of oil a day by 2022/3.[22] Since the oil and gas need to be transported, the Ugandan and Tanzanian governments have signed an agreement with Total, a French oil company, and a Chinese state-owned firm, to build a pipeline 900 miles (1,440 km) long from Kabale to Tanga on the coast of Tanzania, for export to international markets.

We're told that the East African Crude Oil Pipeline (EACOP) and the construction of the refinery will provide 13,000 jobs in construction and 3,000 in operations and maintenance, as well as bringing economic development to the area.[23] We're learning that the pipeline will require the resettlement of hundreds of families, and will run through forest reserves and areas that are especially biodiverse—with elephants, chimpanzees, hippos, and many varieties of birds. Some NGOs calculate that the carbon footprint of the oil when it's burned will be equivalent to Denmark's.[24]

Along with many climate activists in Africa, I've been campaigning, on- and offline, to get EACOP stopped, and for Total, the largest shareholder, to step out of the project. As part of this effort, my colleague Evelyn Acham wrote an op-ed in March 2021 for *Africa News* about what's wrong with the project. With coauthors Charity Migwe from 350Africa.org and Edwin Mumbere of the Center for Citizens Conserving, Evelyn argued that:

> East Africa does not need oil or any fossil fuels to unlock
> its future especially when there are viable, affordable and
> clean alternative sources of energy such as solar and wind,

which are renewable and have better prospects when it comes to long-term job opportunities. East Africa needs to focus on a just transition to renewable energy that guarantees the extensive deployment of millions of clean jobs.[25]

It's true that Uganda has high unemployment and of course an ordinary citizen can ask, with justification, "What will give me enough to eat and survive on? How do you expect me to live sustainably if I have to leave this job?" This is why Evelyn asks why our Government doesn't try and create more green jobs, like in the informal recycling of plastics, which, she reminds us, are the biggest pollutants of Earth and its oceans. She suggests that some of these green jobs "could be more artistic, selling rather than dumping the plastics." They could help reduce poverty and high rates of unemployment, and empower youth, women, and young girls.

I agree. Instead of governments facilitating or subsidizing wealthy corporations' access to petroleum projects, they could incentivize the production of environmentally friendly energy. They could make cleaner, less resource-intensive technologies more affordable and accessible. In the case of cooking stoves, this is not beyond the reach of the Ugandan Government right now. First Climate, a German-based NGO, has provided 450,000 households in Kampala with clean cooking stoves—manufacturing them in Kampala in sufficient bulk to drive down costs and bring the stoves more into line with customers' expectations for price.

There are signs of progress relating to energy. In early 2020, the Government signed an agreement with an energy company from the United Arab Emirates (UAE) to build four solar and two wind farms in northeast and northwestern Uganda. Of course, I welcome this positive step, although by early 2021, it wasn't clear that construction, planned to start by then, had begun. Some observers believe that solar could really take off in Uganda: we have plenty of sun, and right now, about 40 percent of urban residents and 60 percent of people in rural areas don't have access to electricity.[26]

*   *   *

The use of plastic and reducing meat consumption are two other areas where we climate activists are calling for governments to think more creatively and to change the current incentives. Uganda's Government has issued a ban on plastic bags, but loopholes persist. Plastic bottles, meanwhile, are ubiquitous in Uganda, as in many other parts of the world. Ugandan FFF activist Sadrach Nirere is campaigning to #EndPlasticPollution in Kampala and around the country. Plastic, he points out, is not just an *accidental outcome* of people's choices. It's the result of corporate and governmental decisions. Coca-Cola produces four million plastic bottles *per week*, in Uganda alone—a rate of production that no recycling capabilities can possibly handle. As Sadrach says: "Our individual actions are undermined as companies continue to produce plastics with minimal or no responsibility towards the plastic pollution problem they create."[27]

Ninety percent of plastics are made using oil, and globally the fossil fuel industry is looking to transform more oil into plastic and other petrochemicals, notwithstanding the climate crisis unfolding in front of us. Too many governments are helping them, including mine. As Sadrach's website has observed, the Ugandan Government wants to make it easier for plastics manufacturers to access petroleum products made of Ugandan oil. [28]

Another area where the Government could provide direction is around food systems. Agriculture is often the leading cause of deforestation and biodiversity destruction. And larger-scale animal agriculture, which is expanding in Uganda, is particularly carbon- and resource-intensive. Many Ugandans are food insecure right now, and the climate crisis is affecting farmers the most. But the FAO report I wrote about earlier in this chapter illuminated just how difficult it will be for Uganda to produce more meat, as well as enough maize (corn) and soybeans to feed those animals.

Like most people in Uganda—and elsewhere in the world—I had little idea of how food consumption affects the planet. More than one-third of GHGs are generated by the global food system,[29] and the livestock sector alone is responsible for at least 14.5 percent of GHGs.[30] Now that I've been made more aware of the extent of this contribution to climate change and loss of biodiversity, I don't eat meat or other animal-based foods every day. Although I'm not a vegetarian, and before I became an activist I'd never heard of the word *vegan*, I'll sometimes not eat meat for a week or a month. In fact, I shocked my sisters by telling them that if I ever have a wedding it would be vegetarian, since not serving meat at a ceremony is considered rude or just miserly. (I offered a concession when I assured them it wouldn't be vegan!) Still, I've mused about opening a small vegan restaurant in Kampala someday. It might be the first.

Of course, there are other cultural as well as structural obstacles to reducing meat consumption and making greener choices. Meat consumption in Uganda is a way of signifying wealth, as it is in many countries. If you're eating beans or peas you're assumed to be too poor to afford meat. The allure of the Western diet has reached Uganda. Fast-food franchises, such as KFC, are becoming ubiquitous in Kampala and the prices are relatively affordable—making them appealing to students in secondary schools and at universities. There's even a KFC in the vicinity of MUBS, my alma mater, and the main campus of Makerere University. Eating there is seen as modern and kind of fashionable, and students and young professionals often take photos of their meals and post them to social media.

Most urban workplaces don't serve food, which means you have to buy it for yourself at lunchtime, often from restaurants that only or mainly serve meat; chicken is especially popular. If I suggested to people that they eat less meat for environmental reasons and their own health, I'd likely encounter resistance: "You don't expect me to go hungry when I'm at work, do you?"

The Government could encourage restaurants and food stalls to serve more vegetables, including beans that are rich in protein and iron. But there'd likely be pushback. If I mentioned to Uganda's leaders that as a society we should think about reducing meat consumption among middle- and upper-income citizens, for the climate and public health, most would probably say to themselves, "What is she talking about?"

Clockwise from top left: With my siblings and cousins, celebrating my sister Clare finishing primary school; During my first year; With my mom and aunt; With my dad at my second birthday party.

All photos courtesy of the author, unless otherwise noted.

Clockwise from top left: Celebrating my graduation from primary school with high marks; In primary school doing a class presentation during the music, dance, and drama competitions. I am in the middle; Less than a year old, 1997; A glamor shot from my teenage years.

Above left: Striking outside the Ugandan Parliament, January 2019.
Above right: A strike outside the Shell station in
Bugolobi, Kampala, later in 2019

Participating in the Global Climate Strike in Kampala, September 2020.

A climate strike during lockdown, May 2020

Elizabeth Wathuti, Adenike Titilope Oladosu, and me, Ibadan, Nigeria, November 2019

A climate strike at COP 25, December 2019

With Greta Thunberg at COP 25, December 2019

A climate strike in Luzira, Kampala, June 2020. You can see a boda boda motorcycle taxi and matatus behind me.

Cooking stove installation, at Damali Nabagereka Primary School, Wakiso District, April 2021

The Arctic Basecamp climate activist delegation with model-activist Lily Cole (center), Davos, Switzerland, January 2020.

Me, Luisa Neubauer, Greta Thunberg, Isabelle Axelsson, and Loukina Tille, Davos, Switzerland, January 2020.

Participating in the Global Climate Strike in Kampala, September 2020.

7

Speaking Out for Women and Girls

One morning, when I was only a few years old, I went missing from home. My parents looked high and low for me. Eventually, I was found sitting in a classroom at a nearby nursery school, ready to learn. The same thing happened the next day. My father asked the teacher what could be done, since I was too young to be enrolled in the school but I threw a tantrum if I was taken out of the class. The teacher replied that it was illegal for my father to pay to educate an underage child, but agreed that I could stay for free. So, every morning I attended the school, and I was eventually enrolled, a year earlier than I should have been.

I obviously had a hunger for learning, and I'm very fortunate my parents championed my education, particularly as a girl, and worked hard to ensure they had the money for school fees, even in some lean years. My mother stressed that there was no future for me if I didn't study and acquire expertise and skills to be able to earn my own income. To her, and to me, financial independence for women, whether they're married or not, is essential.

My father was equally insistent. His father had done his best to put him and his siblings through school, and he wanted the same for us. Like my mother, he saw how a lack of education limited the prospects for children, especially girls. He wished to empower my two sisters and me to grow up to be strong women who knew their rights and could assert them to attain a better position in society. Like me, he received his degree from MUBS.

My parents' passion for education has extended not only to their five children, but to three of our cousins, whose school fees they contribute to, and whose education and prospects they care about as much as ours. As for my siblings: my sister Clare is at university studying to be a veterinarian while Joan finished high school in 2019 and earned Government sponsorship for university. Paul Christian has passed into the last two years of secondary school, and Trevor is in his sixth year of primary school.

Educating girls is not a high-tech or new idea, and has been a pillar of global development policies for decades. In fact, you'll likely have heard many leaders, women and men, testify to how important it is for girls to be in classrooms on an equal basis with boys. Uganda has mostly reached parity between girls and boys in primary school education. That's an achievement, of course, but many thousands of girls and boys still aren't in classrooms, and many girls, like my mother in her generation and some others in my extended family now, leave school before they complete their secondary education. That means relatively few Ugandan girls will attend university. When I was a student at MUBS, I saw plenty of young women in my classes. Even though there were many women at the university, there are many more who aren't.

Of course, I care about boys' education as well, and with two brothers, I have to. But across sub-Saharan Africa, at least 33 million girls who could be in primary and lower secondary school aren't (equivalent to elementary and middle school and the first two years of high school). More than 50 million girls in the region are missing out on receiving an upper secondary school education (equivalent to the last two years of high school in the US or the sixth form in the UK).[1] Around the world, more than 130 million girls aren't in school and should be.[2] If they had the chance, how many of these young women could be teachers, lawyers, doctors, NGO staffers, members of parliament, or climate scientists?

I think of it like this: girls and women are more than half the world's population. If we are to successfully address the climate crisis, we need women in the rooms where decisions are being made that affect the climate (and almost all decisions now do). Educating girls brings them into those rooms, and expands the number and approaches of possible decision-makers and solutions.

At the moment, neither the access nor the positive outcomes that would result are happening fast enough. This reality is partly a result of girls' disempowerment. I'm certain there are tens of millions of girls—and there are countless across Africa— who'd love to study through high school and even university. But many others are doubtful about their prospects and their own capabilities: *My mom didn't make it even this far in school*, they tell themselves, *so what makes me think I can make it to that level, or even further? You're a rural girl*, the voice whispers, *you probably won't go anywhere in life, even with some education. Why keep going?*

If you don't marry, have a child, and settle for a life of

motherhood on a farm, scrabbling for food and fuel for your family, what options await you? You could move from your village to Kampala and work as a maid in the home of a wealthier family.

Perhaps that's the life of the young woman who's walking by as we're holding our placards at a strike. *What are they doing?* she might ask herself, with a mixture of curiosity and puzzlement. But there's no time to think more; she's likely in a hurry, seeking the fastest ways of securing the household items her employer has requested. So, she gets into a *matatu* (public minibus) and heads back to prepare dinner, or wash the floors, or clean the family's clothes. Honestly, I can't see her paying much attention to us at all.

Could someone like her, let alone a young woman in a village, become a climate activist? How would she have the time? Practically, she'd probably only have a flip phone, not the smartphone so many of us now carry. That would make accessing the Internet difficult and expensive, so she'd be disconnected from what's happening in the wider world. By the time she'd turned twenty, she'd probably be looking for another job. Her employer, worrying that her husband might now look at the young woman with sexual interest, would dismiss her. This is something that happens all the time. Then she would be struggling to find a new source of income and some stability.

Perhaps it's not the worst fate, but is it really all we want for those girls? To me, these are "survival" lives. Would these young women have chosen these futures if other paths had been open to them, like finishing secondary school, maybe completing a university degree, then getting a job and attaining financial independence, or even becoming engaged in activism?

It's a depressing reality that the Covid pandemic has made

situations such as I've described even worse, and in the same parts of the planet where the climate crisis is a daily emergency. Covid and the consequences of climate change have intensified pressures on household incomes across Africa, Latin America, and Asia. School fees, especially for girls, have become a luxury that has to be cut from the family budget, as it was for Hilda Nakabuye. Millions of girls, as well as many boys, may never return to schools once they've fully reopened, and the hard-fought gains in girls' education made in recent decades may be diluted. We may never know for sure how many children and adolescents have been affected, nor be able to add up all the costs of the pandemic to them, to society, and to the climate.

*   *   *

It's because I've benefited from education that I'm so passionate about it. But it also happens that girls' education is a crucial way to address the climate crisis and ensure a more just world. We can't wait for geoengineering innovations, such as capturing CO_2 and other GHGs (greenhouse gases) from the atmosphere—even if we had anything more than an inkling of how viable they are. We need practical and affordable solutions right now. So, why aren't we talking about educating girls and acting on this policy more? Here's why we should.

Project Drawdown, a consortium of researchers and advocates quantifying solutions to the climate crisis, ranked educating girls, together with providing family planning, as the fifth most effective means of reducing GHGs (after onshore wind turbines, utility scale solar photovoltaics, reduced food waste, and plant-rich diets).

Project Drawdown estimates that taking steps toward universal education, as well as investing in family planning in low- and middle-income countries, could result in a massive reduction in GHGs of as much as 85.4 gigatons (expressed as CO_2 equivalent) between 2020 and 2050. That's almost the equivalent of ten years of China's current emissions.[3] Decades of research has found that girls who graduate from high school are healthier, have more economic opportunities, and, crucially in Project Drawdown's calculation, bear fewer children over the course of their lives. They're also more likely to make sure that their kids, including their daughters, are educated too.

I explained why this is the case in an op-ed I wrote for *Wired UK* in January 2021:[4]

Lower fertility can lead to healthier, more secure families, and it reduces emissions well into the future. But, while fertility rates are important, they are far from the only reason why educating girls is important for our climate future. Women are also disproportionately impacted by climate disasters: the UN estimates that 80 percent of people displaced by climate change are female. With the climate crisis, as so often, women's suffering is intensified by the structural gender inequalities that dominate their lives.

Rural women are responsible for most of the childcare. They grow most of the food and harvest the crops; they walk long distances to market, and to collect fresh water and firewood for cooking and heating. All you hear are cries of back pain because of that stress. To assist their mothers, girls may be forced to drop out of school. When even this isn't enough to afford basic necessities, some mothers have to make heartrending decisions, such as sending their children to the city to beg, like the little boy that Kaossara Sani came across on the streets of Lomé, or giving up their daughters for marriage, often to much older men.

In return for the girl, families will receive the traditional "bride price." Usually, it's paid in gifts or money, and sometimes in both. For a poor girl, the bride price might be a few sacks of maize (corn). To a family with other hungry kids, or whose crops have been destroyed by floods, that can make a difference. It's distressing to think that's what a girl forgoing her education is worth.

In some places, early marriage is dismayingly common. Thirty-five percent of girls will be married before they turn eighteen in sub-Saharan African countries.[5] In Uganda, it's more like 40 percent, and 10 percent of girls in my country are married by the time they're fifteen, according to local NGO Uganda for Her.[6] This isn't solely an African problem. In South Asia, UNICEF reports, almost 30 percent of girls will be married by the time they turn eighteen.[7] Many of these young women and girls won't have completed high school and they'll have had little or no education on reproductive health and sexuality. Almost all of them will soon be mothers, a precarious situation they're simply not prepared for.

In Uganda, society places the sole burden on young women not to become pregnant if they're unmarried, and blames them if they do. Dropping out of school may be part of the price they're forced to pay, even while the fathers of their kids are free to continue with their education. At my all-girls boarding school, we had pregnancy checks every six weeks, which entailed a physical examination. Why do we accept and reinforce this double standard?

I know girls in villages in Uganda who've been married at

fifteen or sixteen, or even younger, and by the time they're seventeen they have two kids. They'll probably have many more, whether they really want to or not. (Whenever we'd discuss it, my friends from secondary school and university never expressed a desire for more than two or three kids.) With early pregnancies and a lack of reproductive health care, girls and young women face an increased risk of injury during childbirth, or even death. Although maternal mortality has been declining in Uganda since 2000, it's still scarily high. On average, 375 women will die for every 100,000 births.[8]

Across Africa, and the world, 12 million girls under eighteen years old are married each year, many of them against their will.[9] According to a World Bank study, child marriages cost Africa US$63 billion in human capital.[10] For countries like Uganda, this leads to poverty that persists over generations, and fewer women in leadership roles because they haven't received a university degree, which is often a prerequisite for holding a position in government. It also means early marriages and early pregnancies and an increased risk of complications, permanent injury, and even death in childbirth.

One young woman fighting the stunting of girls' potential is the Zambian Natasha Mwansa. Natasha became an advocate for preventing child marriages and teen pregnancies and supporting reproductive and mental health for adolescents when she was only twelve years old. Natasha is passionate about the need for girls to speak out: "Girls have to demand a space for their voices to be heard," she says. "Therefore, we urge African governments to support and uplift girls, and make a firm decision to end child marriage."[11]

I've often asked myself what my own life would have been like if my parents hadn't committed to educating me and my sisters equally with my brothers. Through my extended family, I get a glimpse. One of my cousins, who lives in a rural part of Uganda, dropped out of school in her mid-teens. She's younger than I am, and has three kids. I'm worried those kids and others she may have are headed for a similar future to hers, because that's all they'll know.

A family friend, a more distant one, also dropped out of school, moved in with a man, and has a couple of children. She's also younger than I am. Not long ago, she came to visit our house. I wasn't there, but my sister Joan told me the story later. Our friend took Joan aside so they could be alone. In a soft but impassioned voice, she urged Joan to finish school. She had been misguided, she said, and hadn't received good advice. As Joan listened, aghast at the pain she could see on her face, our friend added a grim conclusion: her life was now a living hell.

*　　*　　*

Providing girls and women with the means and ability to access education is one thing; *what* we're taught is another. Our schooling isn't preparing us for our climate future. The gravity of what awaits us is hidden from us, not to protect us but to deny us the opportunity to question why we're being educated this way in the first place. School should not simply be about "study, do your exams, and pass." It should give us the tools and information to make decisions about *our* future.

But why delay until secondary school or beyond to confront the crisis? Kids can start learning about climate change in primary school. As already mentioned, Elizabeth Wathuti in Kenya is filling this gap in the curriculum through her Green Generation Initiative. As she explains further: "We provide environmental education and application to make sure that what the children are learning in class, they know how to implement outside." This, she adds, has real-life benefits. "Because sometimes you ask people, 'What did you do with what you learned in class?' It's not so easy for them to learn how to use knowledge to change their society. We fill that gap by training for the practical, teaching them how to make a difference."

Of course, understanding the value of trees and forests is just one aspect of climate education that all students everywhere should learn. Evelyn Acham and I agree that incorporating climate change into every part of the school curriculum would have a major impact. I'd like the climate crisis to be a cornerstone of our education in science, and not just a section in geography or in environmental studies. Here's Evelyn's view:

If the Ministry of Education saw the importance and significance of including climate change education in our curriculum, it could be acted upon. Educating the young about the climate is different from educating older people, because young people are fresh, energetic, open-minded, and ready to learn anything. If we teach them fully about climate change, they can grow into this information and grow up knowing that it's the right thing to do . . . Some of those students starting school are in nine years going to be teenagers. And these teenagers can have a big impact in activism and advising because they've learned so much from school.

Even though I wasn't satisfied with how climate change was taught in my secondary school or at university, being in those classrooms helped me develop critical thinking and other skills to undertake my own research, and realize I had to become an activist. Education is the pathway that provides women with the tools to be more resilient when climate disasters strike—offering them the opportunities for economic empowerment inside their communities and skills to respond to the extreme weather events increasingly affecting many parts of Africa. Still, it's an unfortunate fact that countries across the Global South where the climate crisis is being felt most acutely are the same places where girls are least likely to complete school.

As well as programs to support formal girls' education in Africa and every country where there's a need (and there are many), I'd like more programs for adult women that teach them skills and allow them to believe that even if they've not finished school and have given birth at a young age, they can still be heard. They can still take a stand. They can still have dreams. There are still possibilities.

* * *

In addition to educating girls and embedding climate change in curricula, we need to recognize that gender equality and women's rights are crucial to solving the climate crisis. In addition to ending violence against women and children, and their exploitation, we must support the emergence of more women leaders, popularize the accomplishments of those we have, and challenge some of the social norms that keep girls and women in a less-than-equal position. To me, gender equality starts from societies respecting girls and women and not seeking to control what they can and should do.

As I encountered when I started my climate strikes, the silence of women and girls is encouraged (even ensured) under the guise of maintaining our dignity and self-respect. I've been asked how it is that I not only speak up but do so eloquently. This is a strange question to me, as if it is beyond the ability of a girl to articulate clearly and in full sentences or paragraphs.

Sometimes, I think it's another way of changing the subject from *what* we're saying to *how* or *that* we're saying it. It's a means of policing what is considered "proper" behavior for a woman. This compulsion to control our words and actions can reach ridiculous levels. For instance, in some communities in parts of Africa, women are not allowed to climb trees; but what other solutions do you have when everywhere else is flooded?

Another way to police our words is through trolling us on social media. To speak out on social media is to risk being insulted, ridiculed, and even threatened. This is something many girls and women have been subjected to, and I also have personal experience of.

Of course, social media is an amazing tool for organizing, sharing information, and for encouraging solidarity among climate activists around the world. It's allowed me to amplify my work and push my message, and to do the same for other activists. It has enabled organizations to coordinate their campaigns and their demands, and apply maximum pressure on government and business leaders by letting voices be heard in huge numbers. For those of us with limited access to print or television journalists, or who live in countries without freedom of the press, or who belong to isolated communities, social media has been a vital source of connection and communication.

But social media has also been a magnet for bullying, shaming, disinformation, and even for inciting violence. I'd like to say that the trolls' misogyny, racism, or mockery doesn't affect me. But I won't lie. It can really hurt. Some of the comments posted are truly frightening; particularly those from men who consider me an insult to their very blinkered ideas about what is or is not the appropriate sphere of activity for a woman and especially a young, unmarried one.

I'm not alone. Research by Amnesty International, among others, has found that women, and especially women of color, receive much more harassment and abuse on social media than men. Amnesty's research, published in late 2018 and encompassing tweets received by US and UK women journalists and politicians, also concluded that Black women were targeted most intensely. Black women, according to the research, were 84 percent more likely to be mentioned in "abusive or problematic" tweets than white women.[12]

What's perhaps most distressing is that most of the negative comments I receive come from people in my own country, or in other African countries. As I'd feared before I began my strikes in January 2019, I've been accused in online comments of parading myself in the streets for male attention, of being desperate to snag a husband, or taking drugs (how else to explain my climate strikes?). I've been urged to look for a job, to just get married already and devote myself to cooking and cleaning, and to leave activism to men.

Indeed, when I started my strikes, a friend of mine confided that when she'd posted about it on WhatsApp, her half-sister had replied, "If you were doing that, I'd disown you."

A case in point was in late 2020, when I wrote a letter to then US President-elect Joe Biden and Vice President elect Kamala Harris, asking if they were serious about fixing the climate emergency. (I've put my letter in the appendix.) "All we really want is a livable and healthy planet, an equitable and sustainable present and future," I wrote. "Is that too much to ask? Not to destroy our only home and have a small group of people benefit from our pain and suffering. Let's do all we must to protect our planet and have everybody happy too."

I wrote the letter in my diary, never intending to send it, but keeping it as a record of my thinking at the time. However, when I posted it on social media, the backlash was fierce—especially from young Ugandans, whom I'd expected to be more supportive. Who did I think I was to write to the incoming US president? many commented. Why didn't I stay in my lane? So many people started drafting mocking letters and sharing them with a special hashtag created just for this purpose that some Ugandan media covered it as a story. Mostly, they concentrated not on what I'd written, but on what the trolls had said.

I've tried to comprehend why my fellow citizens and even people my own age would be so negative. Some of my friends have told me it's because I'm doing something different and these people feel threatened or jealous; or they don't understand what climate change is and so they attack it. Others theorize that the attackers think I'm making a lot of money and consider me a stooge of a foreign power, or because I speak out, I must not live in Uganda.

In these circumstances, I block the trolls or try not to respond, because I've found that when I've replied, they've only attacked me more. What I don't understand is their wish not just to annihilate you publicly, but to destroy your inner self, so you crash and decide to give up. I try to focus on the many supportive messages and ignore the few hateful ones, and so protect my mental health from anyone who isn't adding anything positive to my life.

It's never easy, but it's been made more difficult by the fact that, since the pandemic hit in March 2020, most of my activism has had to move online. Because most school students in Uganda don't have access to the Internet or can't afford it, and schools were closed, it was impossible to continue our in-school strikes. So, we tried to make as much noise as possible through social media, to keep people's attention focused on the climate crisis, which would outlast the pandemic. But there were consequences of our inability to engage in physical activism: for example, multibillion-dollar investment deals in fossil fuels could be signed without activists storming any government or corporate offices.

Social media for all its faults has proven necessary as a means of reminding leaders and activists alike that we're still here, like the climate crisis, and to encourage new activists to join us in demanding massive change. Leaders would be very happy if we'd stopped campaigning and fallen silent.

Still, I've learned it's better for my health to ration how often I check my phone to see how many retweets, shares, or comments I've received. Social media can be a lot of work if you give it all of your time, and it can be draining and dismaying to see things you don't expect, like indifference or contempt. It's exhausting worrying about whether you're missing some important piece of information, or failing to retweet or share someone else's message. It can lead to poor eating and disrupted sleep. On several occasions, I've gone to the hospital because of intense migraines, and I can feel pressure behind my eyes because I've been on my phone too long. Now, I try to schedule time to spend on social media at the start and end of the day. I remind myself that social media isn't ending anytime soon.

*   *   *

Although we need to protect our mental and physical health, encouraging women and girls to speak up—whether online or in person—is essential precisely because we're on the front lines of the climate crisis. Ecofeminist Adenike Oladosu describes what's taking place in countries across the Global South: "Women are disproportionately affected by climate change because of their closeness to their environment," she says. "Whenever there is a crisis, they become the first victims: whether displaced by floods, or their farmland being inundated or struck by drought, and the subsequent low yields of their crops." They are, Adenike continues, both the first victims in the climate crisis and the first responders. But they have limited power to advocate for their rights or needs, because so many are in the informal sector of the economy, rather than part of the wage-earning workforce.

That victimization can take several forms. For example, droughts and invasions by locusts in northern Kenya are creating more poverty, food insecurity, and desperation. That, in turn, is leading to more child marriages (so families can receive the "bride price" or dowry) and more instances of female genital mutilation (FGM). Even though FGM is illegal in Kenya, for some communities there, as in other countries in Africa and the Middle East, it's seen as necessary for a girl to have undergone FGM for her to get married.

When women have to spend more time fulfilling basic needs, they have less opportunity to care for their husbands, who sometimes respond with physical violence or abandonment. Men, too, feel the pressure of poor harvests, livestock loss, and job insecurity: in frustration, shame, and a sense of betrayal, they may turn to alcohol and some may become physically abusive.

The United Nations Development Programme has documented that in natural disasters—which we now recognize are really *climate* disasters—women become targets for physical and sexual violence. Some farmers or landowners insist on trading sex with women for food or rent; many victims of that abuse have to sleep on the streets, which can be extremely dangerous. Kids, too, find themselves abandoned to the streets, where they can be preyed upon and exploited. Others have their hopes and dreams dashed. For many, the trauma is something they cannot easily recover from.

Although the vicious cycle of climate change and violence toward women and girls is a dire problem in the Global South, it's not uncommon in the Global North either. The UNFCCC (the UN Framework Convention on Climate Change) itself says that climate change "is recognized as a serious aggravator of gender-based violence," including domestic abuse and discrimination against communities of Indigenous Peoples, and in the context of sexual and reproductive health.[13] It cites a study from Australia that documents a rising incidence of domestic and family violence in rural regions after a series of fires and droughts depressed agricultural income.[14] The disruptions, job losses, illness, stress, and enforced togetherness brought about by the Covid pandemic have increased the instances of gender-based violence. UN Women calls this the "shadow pandemic."[15]

In my own family, we have a relative who's suffered repeated physical abuse from her husband. Learning about this catapulted my two sisters, Joan and Clare, into becoming activists on gender-based violence. They discovered that an estimated one of every three women worldwide, or 736 *million*, have suffered physical and/or sexual violence from an intimate partner, or sexual violence from someone who isn't a partner (not including sexual harassment).[16] As my sisters wrote in a blog post for 1Million Activist Stories, "countries with the highest level of violence towards women have something in common. They have a low rate of women['s] education."

My sisters have now become strikers themselves, both off- and online, on behalf of domestic violence awareness. In their blog, they said their goal is to "create awareness for all the women out there, to encourage them and tell them that they have rights and to fight for them, to stand up against the violence towards women." Joan and Clare are calling for more investment in girls' education and systems that track gender-based violence in homes and schools, and for abusers to be punished, along with stronger laws and better enforcement of them. Their first blog post ends with words I'd like everyone in the world to accept: "Human rights are women's rights and women's rights are human rights."

* * *

To my mind, it's no accident that a wave of young people has swept the world demanding action to address the climate crisis. It's also no accident that young women are leading many of these movements. We've seen what's happening on the ground, we have less access to resources and power, and so we feel more acutely what occurs when the little we have is taken from us—washed away in the rising waters or withering in the unrelenting sun.

It was girls and young women I saw at the forefront of the Fridays For Future movement when I began my research on climate change and climate activism back in 2018. It was really motivating and made it easier for me to decide to become an activist. I could tell myself, *If they can do this, I can too.* If the youth climate strikers had been mostly men, it would have been harder for me to join and to see myself as one of them. I'd have thought, *Well, society gave them that role and so probably that's not a role for me*—especially since, as I've described, in Uganda being an activist isn't widely accepted, and being a woman activist even less so.

For the climate movement, it's important for women of all ages to step into leadership positions at local, national, and international levels. "Women provide solutions and are decision-makers, and we can bring out the best if we are given a role," Adenike says. "We have the solutions at hand to provide people with the means of surviving the climate crisis, and we need to give women the chance to enact them."

Countries led by women are more likely to ratify environmental treaties. A number of female leaders have been urging the world's nations to raise their ambitions and accelerate action on climate change. Among those who inspire me are Christiana Figueres of Costa Rica, who, as head of the UNFCCC, oversaw ratification of the Paris Climate Agreement in 2015; and Amina Mohamed of Nigeria, a United Nations deputy secretary general who chairs the United Nations Sustainable Development Group.

It's thrilling that more countries are embracing the fact that women can be great leaders. So is having a relatively young woman like Jacinda Ardern (who managed the Covid pandemic excellently) serving as prime minister of New Zealand and Alexandria Ocasio-Cortez of New York City becoming the youngest woman ever to join the US Congress. When a girl sees Kamala Harris as a vice president, she can believe she could be a leader too, because now she knows it's possible.

Although the world seems to be more readily embracing women's leadership than it did in the past, we need to pick up the pace, because many women have proven themselves to be better leaders than men. If we were without these and other women in senior positions, climate policy—and just about every other policy—might be worse. For many African women and girls, including me, the late Wangari Maathai of Kenya, who was the first African woman and first environmentalist to win the Nobel Peace Prize, is a powerful role model. She broke many barriers, including being the first woman in east and central Africa to earn a PhD. She was also one of the strongest voices drawing attention to the climate and biodiversity crises and connecting the dots between conflicts and natural resources, both the ones people want, like oil, and those being degraded, like forests and soils.

In my own country, I can look up to women like Winnie Byanyima, who heads UNAIDS, is an under-secretary general of the UN, and served for many years in Uganda's Parliament. I also admire Barbie Kyagulanyi, a businesswoman who's active in empowerment programs for girls and women. She's strong and smart. She's also the wife of Ugandan singer and politician Bobi Wine, who leads the National Unity Platform party and ran for president in the 2021 elections. Too often, societies still only see the role of "wife" when it comes to a well-known man, preferring to focus on him and his achievements rather than hers. Barbie is an amazing person in her own right.

Another Ugandan woman I admire is Her Royal Highness Sylvia Nagginda, the Nnabagereka (Queen) of Buganda, because of the Nnabagereka Development Foundation she founded, which works to empower women, youth, and children. And, of course, I need to add my mom, because of how strong and friendly she is, and the fact that she made sure the boys and girls in our family all received the same full education.

In January 2021, when I watched American poet Amanda Gorman reading her work at the inauguration for US President Joe Biden and Vice President Kamala Harris, I imagined millions of girls around the world picturing themselves speaking someday at a state occasion. I want millions of girls and women to believe they can be anything they want to be, and that they can change the world. If they don't, like a team with half its players sitting out the game, we'll all lose. So will the Earth.

8

Rise Up for Justice

The need for greater climate justice representation among African activists and my wish to attract more of my peers to the movement pushed me to found Youth for Future. This became the Rise Up Movement, which launched its social media platform in January 2020. Rise Up is organized in Uganda by Evelyn Acham, Davis Reuben Sekamwa, Edwin Namakanga, Isaac Ssentumbwe, Nyombi Morris, Joshua Omonuk, my cousin Isabella, who joined me at the first strike, and my sisters Clare and Joan. It also serves as an umbrella group for climate activists in Uganda and across Africa.

Rise Up recognizes that communication and coordination are essential. "We need to work together as Africans, because if we don't step up and we continue to see ourselves as 'backstage,' then we're not going to succeed," is how Adenike Oladosu puts it. "We can't wait for money to step into this room for climate justice. Youth need to step up and step in, and demand a secure future."

Even before my experience of being cropped out of the photo at Davos, I'd noticed the lack of visibility or presence in the global climate movement of people from Africa. But the photo-cropping made me even more determined to support and spotlight African activism, as well as young people throughout the Global South: mainly, but not solely, women. It was clear to me many more of us needed to make ourselves visible and speak out so leaders would *have* to listen to us, whether they were in the Global North or South. It seemed to me that too many leaders were in environments where they were sheltered from the effects of their decision-making. Those of us who had no ability to escape the climate crisis, because it was on our doorstep, had to haunt them: to force them to understand that the consequences of their decisions weren't abstract or insignificant, but in real time and in real life were harming *someone, somewhere.*

The photo-cropping experience also made it crucial that the national and international media move beyond the chosen handful of climate activists they'd usually featured. I and other African and Global South activists wanted them to recognize and include perspectives, stories, and solutions offered by thousands of other young people who could explain their and their nations' climate reality on any platform, at any time. For these activists the climate crisis wasn't a theory; it was part of their daily lives in their communities and countries.

For Evelyn, visibility was essential. As she told me, "I see mostly white activists, from Europe and the US. In Africa, you only hear from a few." The Global North has to widen the frame. As she sees it:

The international community can show support by joining hands with us and amplifying the work that we are doing. We need our work to be shared, talked about, and supported. This can empower people who are already fighting for their environment and give them platforms to speak and opportunities to learn more and get more educated about climate change. Because people listen so much to the international community. It has a lot of power.

Evelyn foresees positive outcomes if there is genuine solidarity in the global climate movement:

The international community can also give us a chance to join their groups, and learn more from them, because I believe they have solutions that we also want to learn. But they need to know us . . . they need to listen to us, too, and the solutions *we* have to offer. Such an orientation comes from knowing and appreciating that everyone has something to say, and not to look down on people from Africa or any of the other continents. They need to value the solutions we give, because they could cause change. We are facing the impacts; we can see what's happening; we're experiencing them.

Providing more visibility and bigger stages for Africans, young or old, is not merely about educating the media on who should be (kept) in a photograph. It's also not about bringing a few of us to a conference and allowing us to speak on a panel. Nor is it about only featuring those who are already popular on social media platforms or have a large audience. That is essentially cosmetic diversity. "Many people are behind the scenes working hard," as Kaossara Sani says. "Maybe people think, 'Hey, she is just the only climate activist from Togo,' but we have people. They may not be on social media. But I can say that they are inspiring me."

Now, I recognize how valuable it is to have role models who

can inspire you and keep you positive: I looked up to Greta Thunberg and others, and I still do; and I hope I fulfill that role for some people. I understand that the media picks individuals rather than mass movements to focus their readers' or viewers' attention: I've benefited from that. I realize that the media tracks who receives more "clicks" or "likes," and that algorithms and editors amplify those preferences in order to sell more advertising. And I'm also aware that organizers promote specific activists at certain events or meetings because they believe the media will come if those activists are present. I've been the beneficiary of those realities too.

We all want to feel welcomed and appreciated by our peers. I can testify how important it's been to feel *seen* in the climate justice movement. But no movement—especially one in which the survival of the planet is at stake—can rely on a handful of "rock stars" or "heroes." Nor should it. We need people of *all* ages and races, with the widest possible range of skills, from every socioeconomic background, and from everywhere on Earth to become involved. Just as there isn't just one activist, or "correct" way to be an activist, so limiting the climate movement to one age group or one form of protest or one part of the globe is to reduce the scope of the potential and power of our collective energy, skills, and voices—and to underestimate the urgent challenges we face.

However, even more than a year after I was cropped out of the photo at Davos, it's hard to escape marginalization. In March 2021, the Berlin Energy Transition Dialogue invited Brianna Fruean from Samoa and me to speak at their virtual conference. Brianna and I were promised five minutes each to make our remarks. In the weeks before the conference, organizers cut our allotted time to four minutes, and then to three and a half. They also insisted on seeing the text of our speeches and repeatedly instructed us not to "name or shame" leaders who'd be taking part.

It's astonishing that institutions want us to be represented so they can claim the virtues of inclusion and diversity, and then determine what we can say, how we can say it, and for how long we can speak, cropping our time to the bare minimum lest we offend anyone. Who should be offended? The leaders who've ignored the crisis or the climate activists who are censored when they speak on behalf of those millions of people whose only offense is to suffer, starve, and die because of the climate emergency: an emergency made worse by the inaction of those leaders' countries?

When I spoke, I went off script. I used my 210 seconds to call the organizers out:

> It is the leaders who have failed us so far—not the young people. It is the leaders who have ignored the scientists and the science. It is the leaders who time and time again have failed to treat the climate crisis like a crisis. This is not "naming and shaming." This is telling the truth. Why are you so afraid of hearing the truth?

*　　*　　*

At the end of May 2020, I watched the video of the white Minneapolis police officer placing his knee on the neck of George Floyd for eight minutes and forty-six seconds, during which Mr. Floyd died of asphyxiation. The incident was covered extensively on Ugandan TV and in the country's print media, as well as online, and it's hard to convey how shocked I was by what I saw, as hundreds of millions of other people were too. The video scared me as much as it angered me. In fact, it tormented me. Whenever I came across it on my phone, I'd freak out and scroll down to avoid watching it again.

The murder of George Floyd and of many other Black men and women made me grasp more fully the sheer amount of anti-Black racism in the world. It wasn't unique to the United States. Black people were subject to it in Europe[1] and Canada,[2] in Latin America[3] and East Asia.[4] Once I looked more closely, with my photo-cropping experience seared into my consciousness, I discovered that bias against Black people was everywhere: in access to health care, in education, in textbooks, in workplaces, in what jobs are available, in criminal justice, in housing, in the media, and also in the climate crisis and the movement that was trying to address it.

While I'd heard of Black Lives Matter and seen some of the movement's Twitter posts before 2020, it was only after George Floyd's death that I got a bigger picture of the movement. I learned about the young activist leaders demanding account-ability and marching for racial justice in the US, and later, in countries around the world. I also saw the call being taken up by activists in the climate movement.

As an act of solidarity, I made BLACK LIVES MATTER, I CAN'T BREATHE, and SILENCE IS CONSENT placards and posted pictures of me holding them on social media. In one of these Instagram posts from the end of May 2020, I wrote: "We all know what is happening right now in regards to Black people. When I say silence is consent, I believe that we all understand what I am talking about. If you keep silent about the killings of Black people, there is a problem." I asked my followers on social media not only to talk about the death of George Floyd and the protests, but to support Black lives and the Black Lives Matter cause. I also shared information about Black Lives Matter marches and did a lot of retweeting of the Movement's posts.

I watched as Black Lives Matter protests quickly spread around the world, including to my continent. Thousands of people in Kenya, Senegal, Nigeria, Liberia, and Uganda, among other countries, took to the streets. They were demanding an end to the racism that had been so brutally reaffirmed by the police murders of George Floyd, Breonna Taylor, and too many others, and demonstrating solidarity with Black Americans. In Kampala that June, an organization called No White Saviors held a Black Lives Matter protest. However, not long after they'd gathered, the demonstrators, many of them expatriates, were arrested. Five were Americans. Police charged them with holding an unlawful assembly and violating Uganda's COVID-19 containment rules. In 2020, police in several African countries enforced Covid restrictions harshly, and even fatally.

Africans also marched to draw attention to the fact that state violence against Black citizens wasn't confined to the US. It was common in Africa, too, and almost always perpetrator and victim were Black. Elections routinely become flashpoints that can result in violent crackdowns by police or security forces. Uganda's in 2021 was no exception.

Black Lives Matter protests in African capital cities sought to draw attention to these forms of police brutality. In Kenya in June 2020, people marched in solidarity with Black Lives Matter in the US and to protest beatings and killings by the police during Covid-imposed curfews. In the capital, Nairobi, one of those killed was a thirteen-year-old boy standing on his family's balcony. Police said he was an unintended victim of a stray bullet from the gun of an officer enforcing the dawn-to-dusk curfew.

Activists marching for Black lives in Kenya, South Africa, and other African countries also called for police and security forces to be "decolonized" of colonial-era attitudes and practices, so that their priority becomes protecting people's rights, not using force to contain them.[5] In Nigeria in October 2020, tens of thousands of people took to the streets for weeks to protest the brutality of the government's Special Anti-Robbery Squad (SARS), and extrajudicial killings carried out by the squad. The authorities violently disrupted the rallies, and a number of protesters were shot. This response drew shocked attention around the world and a statement of solidarity from Black Lives Matter leaders in the US.

"We care about the issues of police brutality, no matter where they're occurring," Opal Tometi, a Black Lives Matter cofounder, said.[6] "People are missing and people have died as a consequence of speaking out. We will not abide it," she added. She and more than sixty other activists, artists, writers, and actors, many of them Black, signed an open letter to Nigeria's president, Muhammadu Buhari, urging that the jailed protesters be released and that Nigerians be allowed to "exercise their constitutional right to protest."[7] Greta signed the letter, which may have helped others understand what we in the youth climate movement knew already: the climate crisis can't be solved without achieving racial justice too.

In my home, we have had many conversations about the murder of George Floyd and the persistence of anti-Black racism and white supremacy. Like billions of people around the world, we were deeply disturbed and angered by what had happened. It was also a grim confirmation of what we already knew about what white people believed they could do to Black people and get away with. It was as if the white police officer we saw in the video thought he had the right to end George Floyd's life in the way he wanted to, even while the whole world was watching.

These discussions also took my family and me back to what had happened in Davos. I remember my parents saying after George Floyd's murder that while they'd thought there was room for white people to make positive changes in their attitudes toward Black people after the photo-cropping incident, now they weren't so sure. A Black man had been murdered by a police officer in broad daylight in the US. Of course, this brought home to all of us that much more terrible things can happen as a result of white supremacy than being cut out of a photo.

It wasn't as if I wasn't aware of the history of racism, in my own country and around the world. In school, we'd all learned about the slave trade and the plunder of African countries by the colonial powers. We'd studied British rule in Uganda and the liberation struggles that led to independence here (in 1962) and in countries across the continent. But the colonial legacy plays out in strange ways. There's still a form of white supremacy that operates in Uganda, and elsewhere I'm sure, because as Africans we have been told to think that white people are above us and we are down below.

I remember being taught in secondary school about the Maji-Maji rebellion against German colonial rule in what's now Tanzania. The "rebellion" (which, of course, should be called a "struggle against colonialism") lasted from 1905 to 1907, and was one of the largest such uprisings in Africa. We learned that Kinjeketile, the rebel leader, had told his fighters that "war medicine" would turn the Germans' bullets to water. It didn't.

That "war medicine" was actually water ("maji" in Kiswahili) mixed with millet seeds and castor oil. As a result of the fighting, and the hunger that followed because crops had been destroyed as well, between 180,000 and 300,000 people died. But the way the teacher presented the story, or the way I understood it, made clear the Germans' superiority in wielding their weapons (guns and bullets) against the rebels' water. The lesson we received from studying the Maji-Maji rebellion was that then, too, white people were way ahead of us.

Today in Uganda, an independent Black majority country, there's a fascination with whiteness and a privileging of it that I've known my whole life. When I was growing up, if kids saw a white person on the street, we'd all get really excited. It was as if we'd seen an angel or something otherworldly—and good. Even now, if you see a Black friend or a Black stranger walking with a white person, it seems almost impossible. People will ask, in disbelief, "How did you become her (or his) friend?" We've also seen this privilege play out in restaurants, shops, and malls in Kampala, when white people may get served ahead of Black Ugandans or be treated in what appears to be a more courteous manner.

The photo-cropping and the Black Lives Matter protests weren't the only incidents in 2020 that made me aware of structural racism. The Covid pandemic also revealed the deep inequities experienced by Black people around the globe. People of African descent, and non-whites in general, were more likely to be exposed to the virus (often due to the "essential" jobs they performed, the housing they lived in, and the health care they didn't have access to). Disproportionately, they were more likely to be infected with Covid and then more likely to become seriously ill or die.[8]

Most of the African continent has been spared the agonizingly high number of infections and deaths that North America, Europe, and Latin America have suffered. Researchers have suggested that one reason for the relatively few deaths in Africa is that most Africans can't afford overseas travel, so the disease hasn't spread as widely. Another explanation offered is that Africans have been exposed to so many infections, such as malaria, tuberculosis, parasitic worms, and respiratory ailments, that their immune response is better able to resist the virus.[9]

Still, even as I write, new strains of the coronavirus are emerging in Africa, and cases are surging, and there's a major disparity in vaccine availability between the Global North and South. This is a further injustice: a by-product of poverty, limited political power, and, in essence, of (most) lives in the Global North being valued more highly than those in the Global South.

And, of course, the inequity between Africa and other continents is felt most keenly in the costs of climate change, and in our efforts to raise the issue. African climate activists are fighting against police brutality and human rights violations meted out by our own governments. We're combating the apathy we perceive among some of our leaders about the climate crisis and about the climate movement. In Uganda, the leadership often prioritizes itself and the needs of the international community and foreign investors ahead of the welfare of its own people and region.

*   *   *

In February 2013, Ella Adoo-Kissi-Debrah had a fatal asthma attack in London after experiencing a seizure, the sort that had required her to be hospitalized twenty-seven times in the previous three years. She was nine years old. Ella's death brought home to me the connection between racial justice and the climate crisis that's one of the least recognized: public health.

I learned about Ella's story in December 2020. That's when the international media reported that a UK court had, for the first time in British history, allowed air pollution to be recorded as the cause of someone's death. The coroner noted that the area of southeast London where Ella lived, Lewisham, had levels of nitrogen dioxide higher than European Union or World Health Organization guidelines.[10] Nitrogen dioxide, which contributes to toxic ground-level ozone, is a by-product of car engines that run on diesel.

We've known for decades the visible damage done to the environment by fossil fuels. We're increasingly familiar with the ever-upward trajectory of parts per million of atmospheric carbon dioxide: reaching 420 in April 2021, a level not seen in recorded history. But much of the climate crisis is invisible. We can't *see* the planet warming or the GHG (greenhouse gas) emissions in the atmosphere. Some people say that if we could—if, for example GHGs were purple—they'd be much harder to ignore and as a consequence we'd have more sustained climate action.

If people understood more about how the climate crisis is affecting their health, I think they'd more fully grasp the urgency of protecting the planet and themselves. The effect of *invisible* particulate matter on our health may be as severe as the visible pollution of oil spills and algal blooms. The particles are so small that they can affect the heart, lungs, and other vital organs, increasing the risk of strokes, heart attacks, and, of course, problems associated with the lungs, such as asthma. My mother suffered from bad asthma when I was younger. I remember the anxiety I felt and the pain in her face as she struggled to breathe. I can only imagine what it must have felt like to Ella and her mother, Rosamund Kissi-Debrah.

Air pollution doesn't only come at a cost to human lives, but to economies in the Global North and South. London has some of the most polluted air in Europe, and the costs to public health, at £10.32 billion (US$14.3 billion), are the highest in Europe.[11] Europe overall doesn't do much better: the public health costs of air pollution across 432 cities on the continent amounted to €1,276 per person (£1,100/US$1,520) or €166 billion per year.[12] In 2016, it was calculated that air pollution lowered Egypt's GDP by 3.58 percent or US$17 billion a year,[13] and Chinese researchers concluded that reducing air pollution would save their country 60 billion yuan (US$9.22 billion) a year in health-care spending.[14] The Centre for Research on Energy and Clean Air has estimated that the cost to public health of air pollution is at least US$8 billion *a day* (or 3.3 percent of global GDP). Indeed, the economic expense in lost work time, medical care, and curtailed lives added up to US$2.9 trillion in 2018 alone.[15]

Of course, the drag on the economy cannot mask the terrible consequences for Ella or anyone else of inhaling so much particulate matter. In Delhi, widely considered to be one of the most polluted cities in the world, more than 50,000 people died in 2020 due to air pollution, according to a report from Greenpeace Southeast Asia.[16] The smoke from burning crop stubble routinely blankets Indian cities, a reality twelve-year-old Indian activist Aarav Seth told me about: "The government has not provided farmers with facilities to decompose or manage the stubble." He adds that deforestation is also making air pollution worse: "Some of these forests are being cut for development, and that is really worsening the conditions of people in India. I want the Government to understand the thin line between development and destruction."

The burning of the Amazon rainforest in 2019 caused almost 2,200 hospitalizations for respiratory diseases in the region.[17] And a report found that smoke from the California wildfires of 2020 had generated even more harmful air pollution than exhaust from cars, and had led to a 10 percent rise in hospital admissions.[18] In terms of air quality, Uganda has nothing to be proud of either. In 2019, the World Health Organization listed Kampala as the fifteenth most polluted city on the planet, with vehicle emissions the main cause. In this analysis, Kampala is more polluted than Karachi in Pakistan, Nagpur in India, and Xianyang in China. (The only other African city on the fifty-strong list was Bamenda, in Cameroon, at number eight.)[19]

Obviously, this issue is personal for me, since I've lived in the city most of my life. Although I'm fortunate that my home is surrounded by trees, when I step out of the gate and onto the street, I can feel the grit in my mouth, taste the thick air in my throat and lungs, and smell the diesel. During some of my strikes, especially around Bugolobi, which is an industrial center, I can see the exhaust fumes linger as the cars pull away. I want to close my nose and stop breathing altogether to avoid inhaling the pollution—especially since leaded petrol is still available for the older cars that most people drive.

Some of the most damning research on fossil fuels and public health is in a report released in February 2021 by Harvard and three British universities. A team of researchers found that more than eight million people were killed by fossil fuels in 2018, much higher than earlier research estimates.[20] Even the researchers were shocked by the results, which they called "astounding." One of them, Eloise Marais, a geographer at University College, London, said, "We are discovering more and more about the impact of this pollution. It's pervasive."[21]

Given the enormous costs to public health and economic activity, along with the tragic loss of individual lives like Ella's, why haven't we dealt with our addiction to fossil fuels in favor of clean, renewable sources of energy?

One reason may be that, like many victims, Ella Adoo-Kissi-Debrah was Black. Neither wealthy nor well-connected, she and her family lived in an economically disadvantaged area of London. Her neighborhood is crisscrossed, as many low-income urban areas are, by highways packed with traffic. It's important that we ask ourselves, if Ella had been rich and white, would she have had to live with and die from such severely polluted air, and would it have taken *seven years* after her death for the coroner to issue his report?

UK climate activist Elijah Mckenzie-Jackson told me that he doesn't think people in the UK took on board the lesson from Ella's death: "She was young, female, Black. The headlines weren't enough," he said. "If we had a middle-class white male who died from air pollution, everyone would know about it."

The reason why I'm writing about Ella, and why the coroner was compelled to hear the case on which he produced his landmark ruling, is that Ella's mother, Rosamund, wasn't silent or resigned. She made extraordinary efforts to make sure her daughter's death had a reason, a *cause*: that some*thing* or some*one* brought it about. She has become a clean air campaigner and has set up a foundation in Ella's name to improve the lives of young people with asthma in South London.

Ella's death—and the deaths of millions of others like

her—are not simply accidents of fate, just as it isn't accidental that she was in the wrong place at the wrong time. The inequalities we see and those we don't—between South and North, wealthy and less wealthy, and people of color and white people—are stark.

Throughout the Global North, Black and other communities of color are more likely to live near sewage treatment plants, landfill sites, and chemical industries; and bus depots and toxic landfills will be located in their neighborhoods.[22] Their residences will be more likely to be situated near slaughterhouses or factory farms that pollute nearby waterways, foul their air, can make them sick, and can cause respiratory diseases.[23] Or they may inhabit low-lying areas, intensifying their exposure to floods, storm surges, and waterborne diseases. Their streets may be less well-maintained or less well-lit, and their apartment buildings may be more cramped or prone to fires.

Their neighborhoods may have fewer trees, and be more subject to the heat-island effect, whereby tarmac and cement absorb and retain the sunlight and warm the surroundings. Here, people may not be able to afford air-conditioners, or they may have jobs that require them to be in the street for long periods of time. As a shocked world witnessed in 2005 when Hurricane Katrina breached the levees and flooded the lowest lying areas of the city of New Orleans, Black neighborhoods can remain cut off, forgotten, and left to fend for themselves by their own government. In Flint, Michigan, also in the US, Black families continue to fight for justice after being exposed to lead in their water supply.[24]

For the workers in the fields, many of them immigrants or subsistence farmers, heat can kill. In the US, a multiyear study by the Centers for Disease Control found that farmworkers were twenty times more likely to die from heat-related conditions than average workers,[25] and since many of them are undocumented or are from Latin America and only speak Spanish, they may not have access to heat-safety information in English, or they may be scared of reporting illness in case they are deported.[26]

Too often, when some people think of environmentalism or climate change, they assume a color-blind or economically neutral perspective, Leah Thomas, a Black writer and intersectional environmental activist living in Los Angeles, told me. Over and over again, Black communities suffer from higher levels of air and water pollution. "Sometimes, when people think about environmentalism, they try to exclude the aspect of race or wealth, and how those things might play a role in who is experiencing environmental injustice." This is a mistake, she says, "because the people who are currently being faced with environmental injustices the most are communities of color, and that's going to continue to happen if we don't address it."

Leah offers a number of potentially transformational ideas for the US Government. In addition to declaring a climate emergency, she suggests establishing a council of youth environmentalists and a council for intersectional environmentalism to work directly with grassroots climate activists. She adds: "I want to see real-time environmental justice legislation that specifically addresses the fact that communities of color are plagued with these environmental issues and makes environmental racism a civil rights violation."

Environmental justice is also at the center of the work of Veronica Mulenga, a climate activist from Zambia. "At first, I didn't know about environmental justice," she told me. "Then while I was doing the research on climate change, I also came across how disproportionately it affects us in the Global South. I was really shocked. We're the ones that are causing and contributing the least to the climate crisis and then we're the ones being affected the most."

Veronica lives with persistent shortages of power. Rainfall in Zambia has decreased, leaving rivers low and dams without enough water volume for the hydroelectric power plants from which Zambia draws 95 percent of its formal energy capacity.[27] "We experience power cuts from eight to fourteen hours or more every single day," she says. People who can afford generators buy them, she adds, but they run on fossil fuels and emit carbon dioxide. Purchasing enough solar panels to power a whole house is expensive. "We're saving to get a solar panel someday," Veronica says of her family. "I would love the international community to help a lot of us here with financial aid and adaptation methods."

* * *

"We cannot eat coal and we cannot drink oil" is a message I've used a lot in my climate strikes and in my speeches. On the one hand, this is an obvious point about our skewed priorities and failure to ensure basic needs throughout the world. But this slogan is also a literal description of what billions of people like Ella Adoo-Kissi-Debrah are doing every day: inhaling and ingesting fossil fuels in particulate matter, poisoned water, and through microplastics—especially if they are poor, or Black, or both.

This reality leads me to the most fundamental questions. When you think about the terrible inequalities, racism, and manifest injustices embedded in the climate crisis, what is the nature of a system that supports these inequities and accepts the devastating consequences of continuing to burn fossil fuels? And how can it be maintained?

It is a system of extractive and unregulated capitalism that privileges the needs and concerns of wealthy countries, and wealthier populations within those countries, who are wealthier precisely because the natural resources of the poorer countries, and their inhabitants, have enriched them. This system would rather destroy the planet for the benefit of the few rather than preserve it for the many. It's founded on greed and exploitation rather than the well-being of the human family and even creation as a whole. It's a system where the costs of the unsustainable lifestyle of the few are borne by the many: in financial terms, in respect to their physical and mental welfare, and in their very future. It enables a privileged minority to be free by constraining possibilities for the rest.

This system can never have enough. It always wants more: more money, more from nature, more from other people. It relies on a lack of consideration for the value of a human life and the planet, and depends on spreading the delusion that everyone can climb to the top of the pole if they follow the rules. But they won't, for the system is destroying the pole itself.

The system is maintained in large part by a fantasy of endless economic growth without cost, blinding itself to the extreme inequalities and ecological collapse. But as we in Africa wake up to each day, and as the rest of the world is coming to learn, the climate crisis is already here and it's come much more quickly and severely than scientists had forecast. The collapse of the Texas electricity grid due to record cold, the ongoing melting of the glaciers in the Himalayas that caused a dam to be washed away, Storm Ciara wreaking havoc across Europe, catastrophic rains in South Sumatra, northern Zimbabwe hit by flooding with 200 homes destroyed, and an avalanche in Afghanistan—all in February 2021 alone[28]—show that we have no time to dream of "net zero" emissions and a post–fossil fuel future arriving by 2050 or 2060. The scientists have told us: We have fewer than ten years, and even 2030 may be too late to ensure a world that's less than 2°C (3.6°F) hotter.

I think of our current situation as a game a chess. Some of us may have been born as pawns, some as knights, some as rooks. Some of us may even be queens, with a much wider range of movement. But none of us has autonomy: we're all pieces on the board of a game that none of us chose to play. We're all playthings for people who have no interest in our well-being. They are simply trying to beat their opponent. We may be sacrificed; we may capture other pieces. But we cannot escape from the board; we cannot change how we move.

What the world is now realizing is that the control the master chess players believe they have is an illusion. Nature is in charge of the game. The grandmasters are, in fact, the king on the chessboard when checkmate is declared: trapped, unable to make another move.

The time for illusions is over. End of game.

9

Forecast: Emergency

Mama Mugerwa sells *gonia* (plantains) in the market in Luzira, where my family used to live. My mother befriended her when she used to shop at that market. What she earns from the *gonia*, Mama Mugerwa uses to support her four young granddaughters. When COVID-19 reached Uganda in March 2020, and the Government locked down the economy to prevent infections spreading, prices for food shot up. Although President Museveni told shop owners they'd be fined if they raised their prices, vendors blamed suppliers and distribution, and food shortages followed almost immediately. Locked down myself, I watched newscasts in which women like Mama Mugerwa were interviewed crying for help. Unable to work (or their husbands having lost their jobs), they couldn't feed their children or grandchildren even two meals a day, let alone the three that I and my family were able to eat.

It was painful hearing these stories and learning that many thousands of people, including children, were afflicted by the ache of empty stomachs, with the end of the pandemic nowhere in sight. I decided I had to do something. I took to social media and set up an international crowdfunding campaign. I raised enough money to buy sacks of *posho* (maize flour) to feed fifty families, including Mama Mugerwa and her granddaughters.

Despite the relief that money provided those families, I recognized those sacks of flour were only a fraction of what is necessary to stop the hunger many poorer Ugandans had to contend with during the first few months of the pandemic. Nor would it solve the problems of poverty and food insecurity, which are long-standing injustices, not only in Uganda but in many countries.

The livelihoods of Mama Mugerwa and hundreds of millions like her can be tipped into desperation at any moment by an outbreak of disease, a flash flood, a landslide, or another disaster. That precariousness is one of the reasons why, in September 2020, I accepted the UN's invitation to be one of seventeen Young Leaders for the Sustainable Development Goals (SDGs). The role of the young leaders—young activists, entrepreneurs, artists, and educators from Pakistan, the US, China, Senegal, Colombia, and other countries—is to expand understanding of the SDGs among young people in our countries and engage them in trying to make the goals a reality.

The 17 Sustainable Development Goals

1: No Poverty: End poverty in all its forms everywhere

2: Zero Hunger: End hunger, achieve food security and improved nutrition, and promote sustainable agriculture

3: Good Health and Well-being: Ensure healthy lives and promote well-being for all at all ages

4: Quality Education: Ensure inclusive and equitable quality education and promote lifelong learning opportunities for all

5: Gender Equality: Achieve gender equality and empower all women and girls

6: Clean Water and Sanitation: Ensure availability and sustainable management of water and sanitation for all

7: Affordable and Clean Energy: Ensure access to affordable, reliable, sustainable, and modern energy for all

8: Decent Work and Economic Growth: Promote sustained, inclusive, and sustainable economic growth, full and productive employment, and decent work for all

9: Industry, Innovation, and Infrastructure: Build resilient infrastructure, promote inclusive and sustainable industrialization, and foster innovation

10: Reduced Inequalities: Reduce inequality within and among countries

11: Sustainable Cities and Communities: Make cities and human settlements inclusive, safe, resilient, and sustainable

12: Responsible Consumption and Production: Ensure sustainable consumption and production patterns

(continued)

13: Climate Action: Take urgent action to combat climate change and its impacts

14: Life Below Water: Conserve and sustainably use the oceans, seas, and marine resources for sustainable development

15: Life on Land: Protect, restore, and promote sustainable use of terrestrial ecosystems, sustainably manage forests, combat desertification, halt and reverse land degradation, and halt biodiversity loss

16: Peace, Justice, and Strong Institutions: Promote peaceful and inclusive societies for sustainable development, provide access to justice for all, and build effective, accountable, and inclusive institutions at all levels

17: Partnerships for the Goals: Strengthen the means of implementation and revitalize the global partnership for sustainable development.[1]

Although the concept of the SDGs may sound dry and technical, the goals provide an invaluable roadmap for improving the lives of people like Mama Mugerwa and the Earth's ecosystems on which we all depend. The goals also offer an opportunity for and a challenge to the world's governments, whether in the Global South or North, to leave behind outmoded and unjust policies and priorities in favor of those that promote equity, justice, and resilience, and are climate-compatible.

The seventeen SDGs, approved by all 193 UN member states and designed to be achieved by all nations by 2030, include 169 specific targets. By design, the goals are interwoven. Taken together, they envision a very different world from the one we live in now: without poverty or hunger; with equality, good health care, and decent education for all; with the sustainable use and conservation of land and marine environments; and with significant action to combat climate change and its impacts.

Achieving the targets for all seventeen SDGs is a tall order, although they seem to have more political momentum behind them than the Millennium Development Goals (MDGs), which preceded the SDGs and weren't fully met. High- and low-income countries report their progress on implementation at an annual meeting, called the High-Level Political Forum, each July at UN headquarters in New York. I imagine and hope that most of the world's nations don't want the embarrassment of falling short on meeting global goals *they themselves set on crucial issues* for humanity and the Earth . . . again.

You probably won't be surprised that the SDG I see underpinning all the others is goal 13: climate action. Every SDG is affected in one way or another by the climate crisis, since if targets for goal 13 are to be met, climate action has to include everyone, reach everyone, and be fair to everyone. It must respect human rights and amplify the voices and realities of people in communities suffering the climate crisis in their daily lives.

This is how I see it.

The year 2020 saw record warmth. In 2019, natural disasters, made more frequent and intense by climate change, affected 91 million people.[2] By 2030, according to World Bank research, the climate crisis will have propelled at least 132 million people into extreme poverty.[3] Yet 2030 is the year when we should be celebrating the SDGs being fulfilled. SDG 10 is reduced inequality within and among countries. But climate change is magnifying the disparities billions of people live in the shadow of each day. And the Covid pandemic has shed light on, and intensified, the privations of so many people in frontline communities.

We won't be able to eradicate poverty, the first SDG, without dealing with the consequences of the climate crisis, which is building a trap, and making inherited poverty more likely for the next generation, and generations beyond that. Millions of families around the world make a living growing crops and selling them in markets, like Mama Mugerwa. When a natural disaster strikes, they're pushed to sell the few assets that survived the catastrophe to pay for food or to secure rough shelter. How can families who lose everything in a few hours escape poverty?

How can we reach goal 2, zero hunger, when hundreds of millions of people already don't have enough nutritious food to eat, and when extreme weather and locust invasions are destroying food crops and scrambling farmers' ability to plan for when to sow and to reap? "Right now, the situation is confusing to farmers," Elizabeth Wathuti explains. "Kenya has been having rains in January, but January is known to be a dry season, without even a drop of rain. Sometimes the crops end up failing. Other times, they end up not having a good harvest." The effects can be felt across the country. "When there's no good harvest, especially for crops like maize [corn], then there's a food crisis," she says.

We won't be able to achieve goal 3, good health and well-being, if we keep burning fossil fuels, which is increasing air pollution in our communities, and which, in turn, is destroying people's health—people like Ella Adoo-Kissi-Debrah.

Goal 4, quality education, is also undermined by the climate crisis. Families who've lost farms, crops, or homes can't afford school costs. Kids can't learn in schools that are flooded, and can't study when schools have to shut for extended periods. In some places in Kenya, Elizabeth explained when I interviewed her in February 2021, the flood waters hadn't receded: "Some schools have still not resumed, because the water levels are too high." In other schools, pupils had to use a boat to reach their school. In a few more years, how many boats will the world's children need?

Girls bear the brunt of these disruptions: pulled out of school or encouraged to drop out, often married young, robbed of their dreams and independence. Up to *ten million* more girls could become child brides by 2030, UNICEF projects: a direct result of the Covid pandemic alone.[4] How will we meet goal 5, gender equality and empowerment of all the world's girls and women, when so many girls aren't completing their schooling, or they're marrying as children and losing opportunities to make decisions and be independent?

What about goal 6, ensuring everyone has clean water and sanitation? At least a third of the world's people, about 2.5 billion of us, don't have a ready source of clean drinking water. Millions also lack proper toilets. Lack of sufficient water risks displacing 700 million people by 2030, according to the UN.[5] Sometimes, there's too much. Rising water levels triggered by climate change are submerging toilets and contaminating water used for drinking, washing, and cooking. This exposes babies and young children to malaria and diarrheal diseases, which can be fatal. Every year, about 400,000 people die of malaria; about two-thirds of them are children younger than five.[6]

I was almost one of them. In my first year of life, I contracted malarial dysentery. When my parents took me to the hospital, the doctors struggled to find a vein through which they could inject fluids to rehydrate me. Six doctors tried without success. My parents became increasingly desperate; my father feared I might die. Eventually, one of the doctors got some drops of water into me through my nose, which helped. I recovered and, thankfully, I haven't been sick like that again.

Goal 13 helps determine whether goals 14 and 15, life below water and life on land, can be met. Each of these three goals are tied together. Forests are "one of our most critical allies in climate action . . . however, they're being lost to development," says Indian climate activist Aarav Seth. "This is hazardous for aquatic life and plants, and wildlife near them." He adds, "I'm sure that if we remain blindfolded, then we will need to pay a higher price for our blindness."

Halting biodiversity loss and deforestation are two of goal 15's targets, and are supposed to be achieved by 2020, not a decade later like most of the other targets. Clearly, progress toward both targets is wildly insufficient. More than 30,000 species are still threatened by the possibility of extinction and forest loss continues at an "alarming rate," according to the UN SDG monitoring office itself, due principally to the expansion of agriculture.[7] Kaime Silvestre, the Brazilian climate activist from the Amazon region, told me: "Millions of people depend on the Amazon and the forest, not just Indigenous Peoples, but traditional communities too. We're facing more fires than ever, which is terrible, especially for the lives of animals and Indigenous communities. Livestock production is also destroying the forests." Kaime is seeking to build global solidarity in the climate movement, including for protection of endangered lands. "The Amazon situation impacts the whole climate and global weather patterns, which means that its destruction has consequences not just to Brazilians, but the entire global population," he says.

Just as goal 13 is fundamental to the achievement of the other goals, so justice must lie at the heart of goal 13. Climate justice means that countries act together, accept their different responsibilities for creating the crisis in the first place, and then apply their resources to address the emergency. Mary Robinson, former president of Ireland and UN High Commissioner for Human Rights, has lent her voice in centering climate justice in the SDGs. Climate justice, she says, "insists on a shift from a discourse on greenhouse gases and melting ice caps into a civil rights movement with the people and communities most vulnerable to climate impacts at its heart."[8]

Earth has a fever, but people have different temperatures, depending on who they are and where they live. We're all sick, but those who are suffering less should help those who are suffering more. As Kaossara Sani says: "We are those who are destroying this planet, and no matter where you come from, Africa or rich countries, Western countries, we are all in this together . . . although maybe in different worlds. We have to clean up the mess together."

"Climate change is happening now and to all of us," UN Secretary General António Guterres has said. "And, as is always the case, the poor and vulnerable are the first to suffer and the worst hit."

Climate justice entails extending the right to life, a viable habitat, and the means to survive beyond your immediate generation. That means a planet where ecosystems and habitats are protected, not destroyed; where the rights and livelihoods of the stewards of natural resources, including Indigenous Peoples, are respected; and where no more coal-fired power plants are built, fossil fuels are left in the ground, and clean energy is universally accessible. Now is the time to protect our life systems.

"It's not enough to take small steps if we want real action," teenage Turkish climate activist Deniz Çevikus told me. "We need brave policies and ambitious targets to make the changes we have to make." We cannot return to what we used to be. Leaders must act like the adults they are.

Otherwise, the SDGs risk being inequitable and unmet, leaving billions of people gasping for clean air, searching for basic necessities like food, water, shelter, and sanitation, and unable to find decent work in a habitable environment. That's why climate justice has to be elevated among high-level champions of the SDGs and throughout the UN system.

*   *   *

In this book, I've used the terms *climate change* and *climate crisis* to describe the consequences of *global warming* or *global heating* of our atmosphere. To reflect the worsening situation over the last thirty years, to emphasize the need for unprecedented action, and to embed the concept of justice within goal 13, we climate activists favor the term *climate emergency*. The word *emergency* accurately describes our predicament and also accurately directs the urgency with which governments should approach any policy or action.

Elizabeth Wathuti puts it this way: "Some nations still seem to be in denial. For them, it's like there's no crisis. Yet we feel it every day. You don't need science or data to know there's a problem." She adds: "Yet people who have all the statistics still seem to be shying away from the real problem."

I agree. If our governments declared a climate emergency,

and not *only* declared it but moved decisively toward the drastic decisions required for their countries and Earth as a whole, it would be much easier for people to understand how serious the issues are. In turn, this would help them see that their governments were serious about addressing them. If governments put in place incentives for systems that made it easier to live, travel, light and heat their homes with green energy, and access climate-friendly foods, their people would be much more likely to embrace these practices.

"I would like my Government to declare a climate emergency," says Leah Namugerwa. "We saw that the Government's COVID-19 message came out very well. It showed me that people can respond when the message is provided to them." Our Government needs to tell Uganda's people, Leah continues, that we are facing a climate crisis that will affect future generations, and everyone, whether they are rich or poor, educated or uneducated.

So, how would I suggest that governments and industry respond to the climate emergency, particularly in Africa? First, as Greta Thunberg has made clear, they must listen to the scientists and follow the data, which is getting stronger and more urgent by the day. Any honest reading of that science must lead all those in power to accept the fundamental injustice that Africa faces: that those most severely affected by the climate crisis *now*, and who will be disproportionately affected in coming decades, live in countries and regions whose contribution to GHG (greenhouse gas) emissions are minimal compared to those of the largest emitters. That means that financial mechanisms, development policies, and economic structures need to confront the historical and ongoing injustice of the exploitation of our nonrenewable resources, and take responsibility for generating real solutions for survival.

Secondly, we need a new definition of development. For many institutions and industries in the Global South, "development" means what governments in the Global North have been practicing for two centuries: fossil fuel–based industrialization at the expense of the air we breathe, the water we drink, the food we eat, and the natural world. Institutions and businesses—whether based in Europe, China, Russia, North America, or elsewhere—continue to promote the opening up of environmentally sensitive areas to extractive industries, discounting their costs on the assumption that the wealth generated will offset any social, environmental, or financial damage associated with that industry.

The promise of jobs and development is a common refrain among fossil fuel companies and other extractive industries, in spite of many decades of experience in Africa and elsewhere that suggests the opposite. Too often, as we've seen in the Congo rainforest, foreign companies in league with national governments have exploited regions with an abundant nonrenewable or rare resource: oil, tropical hardwoods, precious metals, or valuable minerals. The products and the riches they generate have been siphoned off to generate short-term profits for those companies' investors, and shareholders, and their governments, and too often line the pockets of corrupt local officials.

Meanwhile, ordinary people discover that instead of jobs for themselves or their children, or the promised infrastructure, within a short period of time their livelihoods, communities, and natural habitats are degraded. In the best circumstances, they may receive money, but it's nothing compared to what investors will earn. In the worst circumstances, subject to the criminality of speculators, boom-and-bust economic chaos, and similar characteristics of the "resource curse," those communities may disintegrate, and may become overly dependent on that industry or even outside aid for their survival.

How will SDG 8, which calls for decent work and economic growth, and SDG 9, which sets goals for industry, innovation, and infrastructure, be achieved when countries have too often put a priority on extraction and low wages over good, green, and sustainable jobs, industries, and infrastructure that benefit communities and countries and don't despoil the environment?

The tragedy of African and other governments in the Global South mimicking the Global North's pattern of fossil fuel–based development is that much of the wealth of European nations was accrued over the last three centuries at the expense of the Global South. For hundreds of years, Africa was robbed of its riches: our timber, minerals, animals, and people. The enormous diversity of our landscapes, languages, cultures, religions, and histories has been flattened and homogenized into an image of sub-Saharan Africans in particular as victims of poverty, hunger, disease, conflict, and misgovernance. It's not surprising that some African leaders now resist being told by their plunderers that they must avoid further exploiting their natural resources. And that they should invest in less immediately available technologies because the fate of the planet is at stake.

Perhaps this is why African governments are currently doubling down on their commitment to fossil fuels. According to an article in *Forbes* magazine about a report published in *Nature Today*, "by 2030, [Africa's] energy generation capacity could rise from 236 gigawatts to 472 gigawatts, with just 9.6% generated from renewable energy sources not including hydropower.

Fossil fuels, the authors found, would account for 62% of total capacity."[9]

More than 200 power plants are planned for the continent, with the majority of them using coal. Indeed, the *Guardian* reports, "Power ships—vast floating power stations, some burning highly polluting bunker oil—are already moored in Ghana, Sierra Leone and Mozambique."[10] The Russian Government is seeking to construct a one-thousand-mile (more than sixteen hundred kilometers) pipeline carrying gasoline, diesel, and kerosene through the Congo rainforest, from the port of Pointe Noire to Maloukou in the Republic of Congo. This will lead to more deforestation and pollution.

*   *   *

As Davis Reuben Sekamwa, who cofounded the website and podcast 1Million Activist Stories with me, observes, the heavy reliance on fossil fuels is a consequence of Africa's continued wealth of natural resources. Botswana, for example, has 200 *billion* metric tons of coal reserves, and Mozambique estimated its production of coal could exceed 100 million metric tons between 2015 and 2020.[11] This coal needs to be kept in the ground if we are to stand any chance of keeping temperature increase at 2°C (3.6°F), let alone the increasingly out-of-reach 1.5°C (2.7°F).

What is needed are three linked actions.

First, multinational bodies such as the World Bank, the International Monetary Fund, and the European Union have to redirect financing away from extraction and fossil fuels toward renewables and mitigation.

"We will need the international community's financial support for our countries to meet our nationally determined contributions (NDCs) for the Paris Climate Accord," Adenike Oladosu says. "There's a limit to how we can adapt. We can neither isolate nor quarantine from the impacts of climate change. Financing is essential, she adds, "for us to provide the necessary conditions to grow and become climate friendly, while stabilizing ourselves as we are disproportionately affected by the climate crisis." These funds have to be channeled into appropriate industries simultaneously, Adenike argues, with the process being "fast and organized, so the financing doesn't wreak havoc." Most important, she says, is climate governance, which means "transparency in financial mechanisms so that we can generate climate justice."

Climate finance *is* occurring. However, according to the World Bank, even though annual investments in renewables surpassed US$500 billion in 2017 and 2018,[12] this number is nowhere near the US$2.38 trillion *per year* the Intergovernmental Panel on Climate Change (IPCC) estimates needs to be invested in energy (principally energy efficiency and renewables) if the 1.5°C (2.7°F) target is to be met.[13] To get on track to meet the Paris targets, the International Renewable Energy Agency concludes that investment in new installed renewable energy capacity has to reach around US$22.5 trillion by 2050. That means at least a 100 percent increase in the amount that's invested each year now, to approximately US$660 billion.[14] But old thinking and investment patterns haven't faded away. According to a report endorsed by more than 250 organizations in 45 countries, between 2015, when the Paris Agreement was signed, and 2019, 35 global banks invested more than US$2.7 trillion in fossil fuels.[15] Yet the International Energy Agency also estimates that US$300 billion worth of fossil fuel assets might be "stranded" by 2035 (once valuable but not so anymore), as the world increasingly wakes up to the climate crisis and governments adopt stronger climate policies.[16]

Climate finance recognizes that "energy poverty" profoundly limits the possibilities for human well-being, as the evidence of the schools I've helped demonstrates. Thankfully, initiatives like IRENA Sustainable Energy Marketplace, Peace Renewable Energy Credits, Sustainable Energy for All, and the Green and Equitable Recovery Call to Action exist. As argued by the Council on State Fragility, an intergovernmental organization cochaired by former Liberian president Ellen Johnson-Sirleaf, these and other groups need to focus on expanding energy access in states that are most vulnerable to the climate crisis, and to concentrate on renewables.[17]

SDG 7 calls for affordable, reliable, and sustainable energy for all. This is now more possible than ever: the costs of renewable energy production and maintenance are now at parity or even lower than fossil fuels. Such investments exist at scale as well. Benban Solar Park in Egypt, as Davis Reuben Sekamwa notes, has the potential to power a million homes. Likewise, the Garissa Solar Park in Kenya, and projects in Algeria, Morocco (which has the largest solar farm in the world), and South Africa are promising—even though they currently represent only a small percentage of energy production for these countries.

In any event, governments of countries such as mine need to stop relying on ultimately self-destructive fossil fuel–based energy production, and be honest with our people about the future and what we need to survive it. "The Government can put in place strict laws, claiming justice and accountability, and the use of energy-saving technology as a means of mitigating climate change. The installation of solar panels," through the

Vash Green Schools Project, "is a sign of hope and change," says Evelyn Acham kindly.

A sign of hope, it's true. But there's no one magic solution to combating the climate emergency. We need many things: from solar panels and cooking stoves to local resiliency and civic engagement in our communities; from expanded investments in girls' and young women's education to tree-planting, reforestation campaigns, and watershed protection; from the reduction of meat consumption where abundant alternatives are available to more diverse, equitable agricultural systems that can withstand the effects of the climate crisis. We must end fossil-fuel extraction, and shift investment patterns and research and development budgets to find and fund innovative ways to meet the energy needs of 11 billion people.

The second action is to reeducate ourselves about what we mean by "development." To avoid total collapse, we must move away from quarterly growth statements, GDP metrics, endless material possessions, and a throwaway culture, toward supportive life systems, drinkable water, breathable air, and vibrant biodiversity. This entails no more logging or road construction in the forests, shelving oil and gas pipelines, conserving wetlands, and stopping the dredging of rivers and beaches for sand. It means preserving indigenous trees and old-growth forests; halting conversion of land to tree, oil palm, or soybean plantations; and ending land grabbing and corruption. The IPCC's report for Africa offers some other useful suggestions: supporting conservation agriculture, including "conservation tillage, contouring and terracing, and mulching," as well as "farmer-managed natural tree regeneration."

Such an approach, the IPCC observes, is as much a social contract as it is an economic or environmental strategy, involving "peer-to-peer learning, gender-oriented extension and credit and markets." It argues for a "pro-poor adaptation/resilient livelihoods" strategy that offers

> improved social protection, social services, and safety nets; better water and land governance and tenure security over land and vital assets; enhanced water storage, water harvesting, and post-harvest services; strengthened civil society and greater involvement in planning; and more attention to urban and peri-urban areas heavily affected by migration of poor people.

All of these strategies require not a top-down approach, but a horizontal integration of many facets of ordinary life for people in the Global South. It's an approach, the IPCC adds, that places "climate resilience, ecosystem stability, equity, and justice at the center of development efforts."[18]

These strategies rely on good governance. When I consider SDG 16, to achieve peace, justice for all, and effective, accountable, and inclusive institutions, I worry. The effects of the climate emergency are derailing efforts to diffuse conflicts and create more just societies as resource scarcity intensifies. Kenyan environmentalist and Nobel peace laureate Wangari Maathai sought to focus the world's attention on these links. To preempt conflict, she said in a speech she gave at the launch of the UN Human Rights Council in 2006, "We must consciously and deliberately manage resources more sustainably, responsibly, and accountably." And, she added, "we also need to share these resources more equitably both at the national level and at the global level. The only way we can do so is if we practice good governance."[19]

Achieving good governance depends on the third essential ingredient for Africa's future: leadership. Leadership requires facing the truth about the climate emergency, no matter how unpalatable, and being honest about what needs to be done— before catastrophes occur. Elizabeth Wathuti would like to see governments doing a better job preparing for climate calamities. "I feel like there are losses we could avoid and prevent. We could avoid the displacement of people even before it happens." But, she adds, "we've got this culture of acting when it's too late, or acting only when disaster hits."

Leadership also means empowering people to reject fatalism. "Climate change is a manmade issue," Kaossara Sani says. "It's not God, like many people used to think in Africa: 'Maybe God is angry with us.' No." Ghana's Chibeze Ezekiel, who won the 2020 Goldman Environmental Prize, undertook a four-year campaign that led to the cancellation of a China-backed 700-megawatt coal power plant. It would have been the first coal plant in Ghana. After the project was stopped, Ghana's government worked on a renewable energy plan. Would it have changed course if Chibeze had not spoken out?

When people are empowered, they will accept responsibility. We face a major challenge making our cities more sustainable and inclusive (SDG 11) and ensuring that all production and consumption is responsible and sustainable (SDG 12). So many of us are used to living in ways that accelerate the climate crisis, and government policies and business practices encourage us. Most people he talks to admit there's a climate crisis, says Roman Stratkötter, a climate activist in Germany who's been part of the Save Congo Forest campaign. But when he tells them, "Great, now we have to fight it," and that this means not flying "every week to Spain for €50" or "eating meat every day for €1.50," many people hesitate. In theory, everyone is a climate activist, Roman adds, but not everyone wants to act like they are . . . at least not yet.

Leadership also means promoting medium- and long-term benefits over expediency. That might include explaining how installing solar panels or buying a cleaner cooking stove or preserving a wetland may seem expensive at the start, but in the long term it will be cheaper because it saves energy, lowers charcoal use and air pollution, reduces flooding, or stops storm surges. Or it may mean establishing more efficient and cleaner public transportation solutions—such as the Bus Rapid Transportation system in Dar es Salaam, Tanzania—so that fewer people feel the need to use private vehicles or *matatus* (public minibuses) that rely on a diesel engine. Simply telling people that they need to drive less will only lead them to ask, "Do you expect me to walk to work?"

The Global North has an essential role to play too—and a historical and ethical obligation to lead and provide adequate financial resources. Governments, mine among them, need to ensure that in addition to capital transfer for green energy projects, international bodies and richer countries engage in an equitable *technology* transfer (SDG 17). The final SDG is focused on increasing development assistance and more sharing of technologies by wealthy countries with less wealthy ones, which would speed up adoption of greener, more equitable approaches for meeting the SDG targets.

I think we can achieve SDG 17, but we have to be careful. The United Kingdom, for instance, has announced a ban on the production of diesel-engine vehicles by 2030. I applaud this decision, but I worry that all those "old" diesel cars will be exported to countries like Uganda, in effect outsourcing the

UK's carbon emissions and negative health effects. Measures like these would defray the costs and hasten the pace of the UK's transition to zero emissions, but they'd delay *our* transition and make it more difficult, more expensive, and more deadly. Similarly, it makes no sense for countries like the UK to commit to ambitious domestic GHG reductions, but continue to invest in fossil fuel–based industries in the rest of the world.

Just as there's no future in Africa recycling the waste or offsetting the pollution of the industrialized world, we don't want to become the junkyard of the old technologies of an irresponsible Global North. Responsible governance means recognizing that if *any* of us are to address the climate crisis adequately, then we must include *everyone*.

* * *

The central question behind good governance, technology transfer, climate finance, and effective leadership, of course, is one I'm asked a lot: *How do we push governments and leaders to take the action we want to address the climate emergency?* Trust me, if we activists had a switch that would make our leaders do what's appropriate, we'd have flicked it already. But we don't, and our task is to keep speaking up, creating awareness, and demanding that leaders do the right thing.

Of course, speaking out is often seen as a threat by autocratic regimes that don't want their citizens to take control of their own environments, or to fight back against polluters. But we cannot afford to be silent. We aren't (yet) in the decision-making forums. We don't (yet) make the rules or (yet) have the votes to determine whether to continue with fossil-fuel financing or to enact change. This shouldn't disappoint us or leave us feeling we can't achieve anything. If we cannot be in those rooms ourselves, then we can put the right people in them—because we have representatives and leaders we can elect to be there.

Many of those who take part in Fridays For Future and other climate actions are still too young to vote; but they're organizing and making themselves heard in videos, online, in op-eds, on the streets, and in their schools. They are convincing their families, their peers, their teachers, and other adults that just as the time has come to transition to a new kind of energy generation, so a new generation with a new energy is transitioning to power. Sixty percent of Africa's population is under twenty-five years old, as is half of India's. Four out of ten people on Earth are under twenty-five. In my country, more than three-quarters of the population is under thirty.

It is this generation—our generation—upon which the future of my country and our world depends, and it's not a matter of *whether* we take power, but *when*. The gerontocracy that's ruled for too long, with its old ways of doing things, its old energies and old conflicts, and its old assumptions about progress and development, is passing away. It has brought the planet to its knees, and the time has come for all of us on behalf of Earth to rise up. We can't be complacent. Instead, all of us have to organize, share, amplify, and vote the right leaders into power: the ones who understand that the era of indecision and ignoring science is over.

Our house is on fire. There's no more time to waste.

10

What Can I Do?

B *ut what can I do?* you may be asking yourself. You may believe that you're not powerful, or that your voice doesn't matter. Or you may not be sure how or what you can contribute to help resolve the climate crisis. I know how you feel, because I once felt the same way.

In this book, I've written about how the media, education, gender, race, and North–South power relations need to shift. I've joined with other climate activists to call for a dramatic reorientation of the economic and political systems that have landed us on the brink of disaster. But the need for systemic change and new institutions doesn't mean an individual can't make a difference. In fact, I believe the change starts within you and individual efforts are essential. As Greta Thunberg says, "In today's debate climate, a lot of people will not listen to you unless you practice as you preach and live by example."

So, this chapter is for you, from me and many of the other activists featured in this book: ten ways to stand up for what is right and just.

One: Find Your Passion and Love

Being an activist can be exhausting and sometimes dispirit-ing. That's why it's essential to discover a passion and a love. What motivated me was learning how the climate crisis was upending people's lives in my country and was the root of so much suffering. I'd discovered a problem and wanted to find solutions.

What is that light within you, that change you want to con-tribute to? You may love your immediate family, friends, and the people around you. If you care about them, you'll fight for them, even if—or perhaps especially if—you're attacked. You may love animals: the beautiful biodiversity that exists on this planet. You may want to save the elephants, or hear the sound of insects, or contribute to the continuation of life on Earth. You may love forests; you may love streams, lakes, or oceans; you may care for the wetlands and peat bogs, on which our shared future depends. You may love your country, or God's creation; you may feel passionate about lifting up marginalized or dispossessed people; you may want to seek justice for those who are disproportionately affected by the climate crisis. There are so many places where your passion might be found.

Whatever your love and passion may be, find them and hold on tight to them. For make no mistake, climate activism is hard. Your love and passion will help you confront negativ-ity from trolls and naysayers who spend their time criticizing rather than helping, mocking rather than offering solutions. It may still be difficult, but with love and passion you can survive the hurtful and abusive comments and insults, and carry on.

Two: *Educate Yourself*

Every generation of activists stands on the shoulders of those who went before. Even though the climate crisis is a relatively recent phenomenon, generations before us have fought for the planet, animals, and for justice and human rights. These people are an invaluable source of inspiration, knowledge, strategy, and experience. We have to work with the older generations if we want to fight the climate crisis. We've a lot to learn from them, as do they from us. So, as Los Angeles, California–based intersectional environmental activist Leah Thomas suggests, "explore environmental history from different perspectives," and honor the ancestors who fought to make the world a better place.

Nobody expects you to be a climatologist; nor should you feel that you have to know the answer to every question anyone throws at you. Our task as activists is to urge citizens and politicians to listen to the scientists and follow the facts. But that *does* mean that it's important to learn. "Keep updated on the climate issues," recommend Turkish activist Deniz Çevikus and English climate campaigner Elijah Mckenzie-Jackson. "Activism involves reading and informing yourself," says Evelyn Acham, "so that when you are speaking to people you do so based on facts. I have personally made Google my friend," she adds.

Hilda Nakabuye agrees: "Read, research, and understand climate change," she says. "Never stop reading. Be inquisitive, talk to other people in the same boat, and ask them for any advice you may need. Don't lose those contacts!"

Three: Find Your People

The world is more interconnected than ever. Ignore those who want to isolate and belittle you, and join with like-minded citizens. When I first wanted to locate those who were passionate about confronting the climate crisis, I went on social media and found people like Greta Thunberg, Alexandria Villaseñor, and Lilly Platt. I learned from them, and their work and commitment gave me energy and courage too. It still does.

Evelyn Acham also recommends finding someone to look up to, "someone who inspires you, someone who teaches you, someone who will help you stand and grow. You need to identify that one person who you think can listen and help you with your activism work," she says. "It may not necessarily be a fellow activist. It can be your parent, your sibling, or any other person."

Connecting with others allows you to not feel so alone, to learn new ideas and information, and to be inspired. There are so many amazing girls and boys, men and women advocating for climate justice: they feed my soul and strengthen my resolve. You can be an ally and a mentor, a listener and an amplifier.

You also need to find your people in your neighborhood, region, and country: volunteers who'll make the placards, or hold them; people who'll step in to hold a strike if you can't make it; and supporters who'll spread the message. You also need friends who'll care for your mental and physical health: who'll tell you when to rest or step back, who'll listen to you during the dark times, be honest with you, and give you good advice.

Four: Share and Connect

I consider myself an introvert. I sometimes find making friends and starting conversations difficult, and I don't like to be the center of attention. I value my relationships with my family and friends, and I try to keep my personal life private. I'm also straightforward and direct, which is why I want you to take to heart that it's possible to be a private person and to share your vision with others, to maintain your family bonds and inspire your friends and relatives to change, and to be a quiet person and still speak out.

It took me a long time to find the courage to stand outside with my placard, and I still grow nervous about appearing in public. But courage not only inspires others, it fosters self-belief and confidence. And just as fear is contagious, so is courage. Of the six of us who took part in that first strike in January 2019, my cousin Isabella has joined me in the Rise Up Movement, Varak has made a video about girls' rights, and Trevor and Paul Christian are more confident and are proud of belonging to a family of activists. Joan and Clare may not have joined that first strike, but they are now activists in their own right on behalf of women's rights and the climate crisis. Directly or indirectly, you can be a positive influence on others whether they're close to you or far away. You never know who you may be reaching.

So, share your dreams, emotions, and what you've learned. It doesn't matter how big or small your audience or stage is, what matters is the message you're putting across, and finding all possible ways to communicate it. You never know where or how people will begin their journey to activism. It's impossible to tell exactly what will spark someone. There may be setbacks and moments of isolation, but we climate activists are engaged in an awesome endeavor, and increasingly we're not alone.

You can start with your parents, relatives, friends, and colleagues, or you can share your information or passion with the world through social media, as some of my friends have. Our friends and family may not want to hear that message, but they are living in the reality of it, and they need to hear it. But you can communicate in a way that makes them want to join you, rather than walk away or worry about you.

Don't limit the ways you communicate your passion and commitment. I've led strikes, given lectures, visited schools, cleaned up trash in colleges and city streets, marched, staged a sit-in, worked with an NGO, sponsored local initiatives (my project for solar panels and cooking stoves in schools), written letters, lobbied legislators, attended conferences, and formed and joined organizations. You could also cook food for and with others and plant trees. All these actions spread your message, build communities and coalitions, allow you to meet new people, and keep you motivated.

Just as there's no one way to be an activist, there's no right time to begin. "You can start now, you can start tomorrow," advises Evelyn Acham. But start! For, as Roman Stratkötter says, "Every little action matters"—even if it's just sharing a tweet or posting your support. "It's still important to make your voice heard," argues Elijah Mckenzie-Jackson, "because everyone has a voice; it's just whether they can find it within themselves to use it."

Five: Speak Out . . .

Many societies expect girls and women to defer to boys and men, stay quiet and not hold strong opinions, and not make statements on public issues. So, to speak out—and to do so firmly, knowledgeably, and clearly—may be threatening to many men, whether in person or online.

But, if you're a girl or a woman, you can't let that stop you. We have a right to be heard, and we have a lot to offer, especially since we're on the front lines of the climate crisis. Find the courage to share your feelings and ideas—even if you fear ridicule, even if you're afraid, and even if you're told you're wasting your time. When an injustice is perpetrated against you or against someone else, speak out. This enables you to make sure the injustice doesn't slip by unnoticed. It allows you to explain why the injustice is wrong, start a broader conversation around that injustice, and then perhaps rectify that injustice.

Six: . . . Listen, Too

Listening is the counterpart of speaking out. Women and girls need to be listened to by boys and men; we need our ideas honored as our own and our experiences validated. Although it's important that activists like me speak up for those people and communities most affected, it's even more important that we give *them* opportunities to explain the challenges they're facing, present solutions, and request assistance.

Listening extends to those in the Global North. We need the industrialized world to stop mouthing platitudes, or inviting us to their conferences just so they can feel better about themselves or promote a kind of cosmetic diversity. The Global North needs to hear *our* reality and support *our* solutions, and then act on them.

If you're from the Global North, familiarize yourself with what's occurring in Africa, such as the floods in Uganda, the destruction of the Congo rainforest, and the shrinking of Lake

Chad. Pay attention to the voices you've heard in this book, and elsewhere in the Global South, and really engage with what they are telling you. If you can, support us with funds so we can continue our work. But also support us with your voice. We need you to speak out on our behalf to your leaders, governments, and industries that continue to invest in and promote fossil fuels, which are destroying our livelihoods, environments, and futures.

We also need to be listened to and our voices amplified because many of us live under the threat of political and/or social violence, and we need you to be our voices if we're ignored by the international community or silenced by the state.

This is an ongoing problem. In February 2021, Indian Fridays For Future activist Disha Ravi was arrested for "defaming India." The Indian police accused her of creating an informational "toolkit" detailing how to support farmers protesting the Indian government's decision to change laws in ways that are seen as favoring big agriculture at the expense of family farmers. Disha's arrest came after Greta tweeted about the toolkit to her followers.

In December 2020, Russian FFF (Fridays For Future) activist Arshak Makichyan was imprisoned "for staging an unauthorized protest" on the climate crisis. My friend Sasha Shugai told the media that she'd been threatened by Russian authorities and the police, and intimidated by teachers from striking. Arshak used his time in prison to educate the police and other inmates about global heating. My own Ugandan climate colleagues have been arrested and Nyombi Morris and Leah Namugerwa had their Twitter accounts suspended in October 2020 after they'd protested against the cutting down of parts of Bugoma Forest.

Japanese climate activist Makoto Sato had her tweets restricted for calling out Japanese banks investing in fossil fuels.

As these examples show, the response of authorities to the climate crisis can be to silence the messengers for calling out their failures to act. This is their message: "If you call for a viable and livable future for humanity, we will shut you down." That's why our message must be: "We will not be shut down. Instead, we will continue to spread a message of hope and action, and in the process reveal the bankruptcy of your cynical disregard for the truth. We will be heard!"

No matter how law-abiding we are, however, we may be marginalized, politically powerless, or absent from discussions. If our stories are amplified, then we'll be more successful. If you live in a country that protects your freedoms, use them to support those that don't.

We also need to listen to each other. None of us is perfect; none of us knows everything; we could all learn more and learn from one another. I've discovered that listening is as important as speaking out. I've also learned that the more voices you listen to, the more effective you'll be as a speaker.

Seven: Be Creative and Take Care of Yourself

There are many activities you can engage in to communicate your message creatively. The activists I met from the Cayman Islands made art and took photos for climate awareness. Elijah Mckenzie-Jackson also uses the arts in his outreach. "You can do so many things," Elijah says: "sing, dance, act, write a drama, do science." You can post an image online, hang a sign out of your window, host a live event on social media, organize a seminar or webinar, draw, give a speech, and demonstrate your talents. There's no one or "right" way to advocate for climate justice: we need all of our skills, our imaginations, and our energy in the way we can best express them.

Leah Thomas is succinct: "Find out what you're good at and use it for the movement." No one can be a bystander; we all must do what we can. Don't worry if you don't consider yourself a very creative person: creativity takes many forms! Here's an example. As we prepared for Earth Day in 2020, which had to be mostly online because of Covid restrictions, we decided on a plan. Each of us would hold a placard that would be prefaced by text that said, "Climate change is . . ." Activists would supply the other word—such as "dangerous" or "here." We took photos and a video of each placard and turned them into media collages. We thought this was more engaging than photos and videos featuring only one activist and the complete message. You can view them online and be the judge of whether our creativity was effective.

As well as tapping into your creativity, it's important to give yourself downtime. "Don't neglect your mental or physical health," Elijah Mckenzie-Jackson advises, "pursue your own hobbies and activities. If you cannot sustain yourself, how can you fight for a sustainable planet?"

If you take your ego out of the equation, which is what Deniz Çevikus suggests, you can more easily deal with the setbacks and negativity, because it isn't about *your* failure or *you* at all. Elizabeth Wathuti recommends being "proactive as opposed to being reactive. Sometimes, we feel so overwhelmed by the complexity of the issues surrounding us," she says, that we become "angry, anxious, or sad." By turning these emotions into action, she adds, "we begin to find and act on solutions."

Eight: Be the Change You Want to See in the World

Mahatma Gandhi's message, "If we could change ourselves, the tendencies in the world would also change," remains as relevant today as when he wrote it. Beyond what we can do individually, we of course need systemic change. Systemic change must elevate all communities, rural and urban, and challenge what's been accepted as business as usual.

So, according to Kaime Silvestre, "to make a difference in the longer term," it's "essential that we are engaged in formal political processes to be able to develop policies which will guide our societies. We young people are often overlooked in political processes, but we are many and we must support and vote for those candidates who prioritize the climate agenda."

So, vote for or campaign for candidates at every level of government whom you believe in, and who agree that the climate emergency is a priority. Run for office yourself. Organize a petition drive. If the international community or your national government seem out of reach, there are many levels of governance that may be more receptive: from your peers in your school to the school board; from your local town or city councilors to mayors and regional representatives. It can be much easier to be an activist where you live. As Leah Thomas advises, "Find out the environmental issues facing your local community" and get involved in whatever way makes sense for you.

And there are lots of things you can do in your ordinary life to bring about change. Speak with your priest, imam, or rabbi about the climate crisis; ask to address the congregation. Engage with your civic societies, such as the Rotarians, or the businesses who can be encouraged to promote the public good. Show up at community meetings. If this engagement works at the local level, you may be able to influence national, regional, and even international bodies to take up your efforts.

Then there's money. Use the power of your wallet. Buy from companies and individuals doing the right thing; boycott those that aren't. Disinvest yourself, your family, and your company from fossil fuel, big agriculture, and other extractive industries. Municipalities, pension plans, individual portfolios, corporations, religious institutions, universities, and governments at every level put their patrons' or clients' or citizens' money into stocks, shares, and other investment products. Start or join a campaign to shift that investment away from industries that are destroying the planet to those that won't.

At the same time, we need to encourage positive investment: investment that puts ordinary people and ecosystems first. More and more portfolios and companies are engaging with a triple bottom line: social, environmental, and financial. Those with means can move their money from banks that support fossil fuels and other climate-damaging industries to those that don't. They can switch their financial portfolios to those that invest in clean, renewable energy.

If it's possible, install your own solar panels where you live or work, and ask your municipality to do so on public buildings like schools, offices, and hospitals. You can also encourage wider use of other renewable sources of energy, like wind. Call on your town or city to offer more public transportation, more and safer bicycle lanes, and more green public spaces. Drive less whenever you can, or don't drive at all. Campaign for a ban on plastic bags and a tax on plastic bottles, and push for discounts on reusable bags and recycling. Work with the local government and businesses to host cleanup days to clear trash from streets, from trees, from parks, along waterways, and beautify your living environment.

These actions benefit us as well as the natural world and improve our relationship with it. As Cristian Martelo, a climate activist from Santa Marta, Colombia, says: "I hope to see the next generation of children walking in the parks and living alongside animals and nature. I hope that someday, animals will see us as friends, not as enemies."

If you work in an office, think before you print, buy recycled paper and print double-sided. Consider ways to cut down on using plastic and wasting food. Advocate for and join efforts to increase everyone's access to healthier, more nutritious, and a wider range of foods. Support a local farm, or family farms, or a food cooperative to lower your "foodprint" and to keep money circulating in the community. Grow food in an allotment or yard, or on your rooftop or windowsill. Share it with kids in school or needy neighbors. Suggest to your family or take it upon yourself to eat less meat or a more plant-rich diet, which Project Drawdown concluded is one of the top five most powerful ways to address global heating. Think about making meals at celebrations or special occasions greener and more sustainable.

And when it comes to clothes, consider how many clothes are in your closet and whether you really need to buy everything brand-new. The global fashion industry is responsible for 10 percent of annual carbon emissions. Many people believe that what looks fashionable and "first class" has to be new. But you can actually look good in secondhand clothes. I often buy them, and I can rock a secondhand blouse or dress for an interview or even a birthday party. My friends will compliment me on what I'm wearing, and when I tell them it's secondhand, almost always the next thing they say is, "Take me to that place! I want to see those clothes."

To be honest, usually my friends aren't compelled by the fact that these clothes are sustainable; it's that they look good and are cheap, so they can save money if they buy them. Not only are these upcycled clothes being kept out of a landfill, but I'm showing my friends that if I can do this, then they can do it as well. And they can be happy and satisfied.

All the ideas I've suggested are ways to advocate for and live a more sustainable lifestyle, and can become part of your ordinary life. Think of Aarav Seth's three "A"s: become aware, amplify, and act. If you use the three "A"s and make this way of interacting with the Earth normal on a local level, then larger change is more possible, because social resistance becomes less.

Nine: Think Globally and Intersectionally

As Aarav Seth notes, "climate activists need to see beyond the borders." Climate change doesn't respect national boundaries, which is why we need to be in solidarity with others. Thinking globally could mean supporting activist causes all around the planet. It could also mean acknowledging your privilege as someone who has benefited from a colonial or imperial past, and continues to do so. That could mean decolonizing culture, as Cristian Martelo suggests, "so that ancestral activities can be carried out to connect with Mother Earth and thus help her recover."

As I hope I've shown in the book, the climate issue affects every other social justice issue. So, as Kaime Silvestre points out, "the fight for climate justice is a fight for justice itself, and must include the fight for social justice and racial justice. Our system has many inequalities. There is no point leaving others behind. We must act collectively and find solutions that truly aim for justice and resilience for all."

Ten: Believe

Change *is* already happening. My children and grandchildren will grow up on a radically different planet. Their realities and the choices they'll be able to make will be very different from ours—and, unless we act now, their options are likely to be much worse and fewer. The question is not whether we act, but *how*. We're fighting for our lives and for the lives to come. Don't be deterred. Believe in yourself and your vision.

Elizabeth Wathuti puts it like this: "Never be afraid to use your voice or stand up for what is right even if you are standing alone." Ignore the trolls and negativity. "Never give up," says Hilda Nakabuye, "however bad the situation may look."

As Greta Thunberg has said, "No one is too small to make a difference." Elizabeth says it another way: "No one should ever feel like their action or whatever they are trying to do is too small . . . We can't hold our hands and watch our planet go down the drain. We have to step up and do whatever it takes to secure a livable world in a safe future for us all and the generations to come after us."

Even though we are individuals, Elizabeth says, "Collectively we are a force, because when all those individual acts are brought together, it's a huge impact." As young people, we, she adds, are "at the center of this problem, being the ones who have to live longer with the consequences." But, she continues, "we cannot avoid seeing ourselves as the solution providers. We have the ideas, the skills, the energy, and the time."

So, start your project or organization, hold your strike, share your idea, and believe in your power. For I believe in you. When I began, I never imagined my activism would extend beyond Kampala or the borders of my country. I never thought I'd attend a UN climate summit or that people would see what I was doing and be inspired to strike in their own countries. What's happened to me, in such a short time, can motivate you, however old you are or wherever you live, to step up and reach out to other people and do all you can. It doesn't matter where you start, or how: what matters is that you begin and do something. The belief lies in you—in your right to be heard, in the power of your voice, and in the strength of your conviction that this *has* to be the way forward.

The climate crisis crystallizes these facts: *We have to save the world. We have to change it, and ourselves. It's not too late.*

Acknowledgments

I'd like to express my gratitude to the many people on several continents who made this book possible and guided me on the journey. First, there is Johanna Langmaack at Rohwalt in Hamburg, Germany, who suggested that my activism could take the form of a book. Second, Mia MacDonald of the environmental action tank Brighter Green and Martin Rowe of Lantern Publishing & Media, both in New York City, whose knowledge, commitment, and skill helped me tell my story, illuminate my experiences, and explore my vision of climate justice in these pages. They brought energy, speed, and expertise to the process.

I also want to thank Carrie Plitt, my dedicated agent at Felicity Bryan Associates in the UK, and Zoe Pagnamenta at The Zoe Pagnamenta Agency in the US, for their support and keen understanding of the publishing process. I am also grateful to my thoughtful and creative editors, Carole Tonkinson at Pan Macmillan, and Rakia Clark at HMH, as well as to Hockley Spare, Jodie Mullish, Jess Duffy, and everyone at Pan Macmillan in the UK, and Taryn Roeder, Elizabeth Anderson, and everyone at HMH in the US.

I know that I am just one among many climate activists across the world, and I'd also like to say how honored I am to be in the company of so many dedicated men and women. I interviewed a number of them for *A Bigger Picture*, and their wisdom, insights, and varied approaches to activism have enriched the book. Thank you to Deniz Çevikus, Cristian Martelo, Elijah Mckenzie-Jackson, Veronica Mulenga, Adenike Titilope Oladosu, Kaossara Sani, Aarav Seth, Kaime Silvestre, Roman Stratkötter, Leah Thomas, and Elizabeth Wathuti.

I also get encouragement, ideas, and much support from many other climate activists whom I write about in *A Bigger Picture* and many I've had the good fortune to have met—whether online or in person. So, a big, big thank you to: Isabelle Axelsson, Xiye Bastida, Sascha Blidorf, Connor Childs, Brianna Fruean, Hindou Oumarou Ibrahim, Eva Jones, Licypriya Kangujam, Elizabeth Gulugulu Machache, Jamie Margolin, Steff Mcdermot, Ndoni Mcunu, Ayakha Melithafa, Makenna Muigai, Luisa Neubauer, Sasha Shugai, Loukina Tille, Amelia "Lia" Tuifua, Alexandria Villaseñor, Olivia Zimmer, and Remy Zahiga.

I particularly want to thank four extraordinary women who have really inspired me: Christiana Figueres, Jane Fonda, Wangari Maathai, and Greta Thunberg.

So much of what you read about in this book would not be possible without my fellow Ugandan climate activists. For your encouragement, solidarity, and commitment, thank you to Rebecca Abitimo, Evelyn Acham, Nyombi Morris, Hilda Flavia Nakabuye, Edwin Namakanga, Leah Namugerwa, Sadrach Nirere, Ayebale Paphrus, Davis Reuben Sekamwa, Elton John Sekandi, and all the others from Fridays For Future Uganda and the Rise Up Movement who have linked arms with me.

That my message has been communicated beyond Uganda is also possible only because of the incredible work of many international friends. So, thank you to Callum Grieve and Connor Turner, for their wise counsel; Tim Reutemann, for his trust and confidence in helping launch the Vash Green Schools Initiative; my colleagues at the United Nations promoting the Sustainable Development Goals; and the many NGOs who have provided me with a platform, including Arctic Basecamp, E-E-T, Fridays For Future, Greenpeace, the Desmond & Leah Tutu Legacy Foundation, Rotary International, the Wangari Maathai Foundation, the World Resources Institute, and many others.

I am also indebted to two of my teachers from my early years of education: Teacher Fred from primary school helped me especially with mathematics, which wasn't easy for me. But he was patient and understood how to get the best out of each student. Teacher Mary, at my nursery school, encouraged me to learn from an early age and told my parents that I could keep attending class, even though I wasn't old enough to be there! I will never forget their kindness, patience, and their belief in me; I was thrilled that they could join me and my family and friends at my university graduation in 2019. I'm also grateful to other teachers in secondary school and at university who helped me learn skills that I didn't think I could, like public speaking. They pushed me and other students to give oral presentations as well as reports to the whole class on group projects we were undertaking. The fact that they wouldn't let me run away from public speaking has proved very useful to my climate activism.

I also want to express my gratitude to the scientists who are giving all of us the facts about climate change.

A huge thank you to my parents and siblings, and my

friends, who have been there all along, always supporting me and never running away.

Finally, my faith in God has sustained me, as has my pastor, Apostle Grace Lubega, who has taught me so much about the Word of God. Activism can be very hard and prayer and attending services (or, in Covid times, watching online) have been extremely important sources of love, grace, and support. My faith offers guidance so I can persevere and reminds me to love everyone—and this has aided greatly in my continuing to speak up for the millions of people in Uganda and around the world who are living the climate emergency right now.

Appendix 1 – My Letter to Joe Biden and Kamala Harris

Dear Madam Kamala Harris and Mr. Joe Biden

I am not really sure whether you will read this but I hope that you do. My name is Vanessa Nakate. I am 23 years old. Congratulations to both of you. I am a climate activist from Uganda who is worried about the present and the future. Are you going to do everything you must to fight the climate crisis? I ask because I really need to know. Climate change is affecting many people's livelihoods in my country, especially children, girls, and women. Are you on our side? I would like to know. All we really want is a livable and healthy planet, an equitable and sustainable present and future. Is that too much to ask? Not to destroy our only home and have a small group of people benefit from our pain and suffering. Let's do all we must to protect our planet and have everybody happy too.

Vanessa Nakate
Uganda

P.S. Please write back.

Appendix 2 – Resources

As *A Bigger Picture* demonstrates, my voice is one among many youth climate activists across the world who are calling for massive, systemic, and rapid change. Below, I've listed social media handles and websites for the organizations and activists I've written about in *A Bigger Picture*. I've learned so much from them (along with others), and I would urge you to follow them on social media, visit their websites, support their campaigns, and amplify their voices. Of course, this list only represents some of the organizations and people standing up for equity, justice, and genuine sustainability.

I've also included other books by or featuring climate activists, and some slogans from my strikes as well as those of other climate strikers, which you can use or adapt. In addition, I've provided selected existing social media hashtags so you can connect with this global movement for change—and become, like me, and millions of others around the world, a climate activist.

T: Twitter; IG: Instagram: F: Facebook; W: Web

My Projects

Green Schools Initiative (W: www.gofundme.com/f/green-
   schools-with-vash)
1Million Activist Stories (T: 1MillionActiv1; IG:
   amillionactiviststories; W: riseupmovementafrica.org)
Rise Up Movement (T: TheRiseMovem1; IG:
   riseupmovement1)
Vanessa Nakate (T: vanessa_vash; IG: vanessanakate1)

Organizations

350.org (T: 350; IG: 350.org; W: 350.org)
350 Africa (T: 350Africa; IG: 350africa; W: 350africa.org)
Act on Sahel (T: ActonSahel; IG: #actonsahel; F: ActOnSahel
   W: actonsahel.net)
Arctic Angels (T: GCArcticAngels; IG: gcarcticangels;
   W: globalchoices.org/arctic-angels)
Arctic Basecamp (T: ArcticBasecamp; IG: arctic.basecamp;
   W: arcticbasecamp.org)
Avaaz (T: Avaaz; IG: avaaz_org; F: Avaaz; W: avaaz.org/
   page/en)
Black Lives Matter (IG: blklivesmatter; W: blacklivesmatter.org)
Congo Enviro Voice (T: CongoEnviroVoice;
   W: congoenvirovoice.wixsite.com/congobasin)
Earth Uprising (T: Earth_Uprising; IG: earth_uprising;
   W: earthuprising.org)
EET: Eleven-Eleven-Twelve Foundation (W: eetfoundation.
   org)
Fridays For Future (T: Fridays4future; IG: fridaysforfuture;
   W: fridaysforfuture.org)

Green Generation Initiative (T: Green Generation Initiative
   (GGI); IG: ggi_kenya; F: GGI.Kenya;
   W: greengenerationinitiative.org)
Greenpeace International (T: Greenpeace; IG: greenpeace;
   F: greenpeace.international; W: greenpeace.org)
LevelUpWomen (T: LevelUpWomen) #AfricaOptimism
Uganda For Her (IG: uganda4her; F: Uganda4Her;
   W: uganda4her.org)
UNICEF (T: UNICEF; IG: unicef;
   W: unicef.org)
UNFCCC (T: UNFCCC; IG: unclimatechange;
   W: unfccc.int)
UN Women (T: UN_Women; IG: unwomen;
   W: unwomen.org/en)
Wangari Maathai Foundation (T: WangariMaathai;
   IG: wangari_maathai; W: wangarimaathai.org)
Zero Hour (T: thisiszerohour; IG: thisiszerohour;
   W: thisiszerohour.org)

Climate Activists Interviewed for This Book

Evelyn Acham (T: eve_chantel; IG: evechantelle; F: Evelyn
Acham) is a passionate climate justice activist from Uganda.
She organizes climate strikes with Rise Up Movement where
she works as a national coordinator. She's part of Fridays For
Future, the international movement of school students striking
for bold climate action. She is also an Arctic Angel for Global
Choices, a youth-led intergenerational action network. She
graduated with a Bachelor of Science degree in Land Economics from Makerere University.

Deniz Çevikus (T: CevikusHB; IG: deniz.cevikus; F: deniz4future; W: denizcevikus.com) is a thirteen-year-old climate activist from Istanbul, Turkey. She began researching the climate crisis in 2018, influenced by Greta Thunberg. In March 2019, she decided to join the Fridays For Future movement, and has been school striking for climate every Friday since then. She's extremely fond of animals and rescued and adopted a tabby cat off the streets.

Cristian Esteban Martelo Ramírez (T: martelocris; IG: martelocris; F: Cristian Martelo Ramirez) is a Colombian environmental activist, focused on the problems of global warming and how it influences the ocean.

Elijah Mckenzie-Jackson (T: Elijahmckenzee; IG: ElijahMckenzieJackson) is a seventeen-year-old climate justice activist based in London, England. Elijah fights for environmental equity, social liberation, and decolonization, as the root of the climate crisis stems from decades of racism and exploitation of the natural world belonging to marginalized communities.

Veronica Mulenga (IG: earth____warrior) is a climate and environmental justice activist from Zambia.

Hilda Flavia Nakabuye (T: NakabuyeHilda; IG: nakabuye hildaflavia; F: Nakabuye Hilda Flavia) is a Ugandan climate and environmental rights activist. The founder of Fridays For Future Uganda, she leads lakeshore cleanup activity on Lake Victoria to preserve water resources and beat plastic pollution. Her passion for nature drives her to create change in her community and globally. She has mobilized and inspired many people to join the global fight for climate justice.

Leah Namugerwa (T: NamugerwaLeah; IG: namugerwaleah; F: namugerwa.leah.3) is a seventeen-year-old Ugandan climate and children's rights activist, and team leader of Fridays For Future and the Save Bugoma Forest campaign.

Adenike Titilope Oladosu (T: the_ecofeminist; IG: an_ecofeminist; W: womenandcrisis.com) is a first-class graduate of Agricultural Economics, an ecofeminist, and the first Fridays For Future climate striker in Africa. She specializes in peace, security, and equality in Africa, especially the Lake Chad region. Adenike is the founder of ILeadClimate and has showcased her climate action in international forums. She has been awarded the highest human rights award by Amnesty International Nigeria for her fight for climate justice.

Kaossara Sani (T: KaoHua3; IG: kaohua3; W: kaossarasani.com) is a peace and climate activist from Togo. She is founder of the Africa Optimism Initiative and cofounder of the Act on Sahel movement.

Aarav Seth (T: AaravSeth; IG: aaravseth_; F: aarav.seth.5454; W: aaravseth.wordpress.com) is a twelve-year-old student activist from Delhi, India. He is the founder of Sunday4 SecuredFuture (which encourages young climate activists to take at least one climate action each Sunday), Helping Hand (which encourages people to donate clothes, books, and shoes to underprivileged children), and She Hygiene (which distributes sanitary pads to girls who cannot afford to buy this basic necessity). He has started a podcast series #RingTheBell (https://anchor.fm/aarav-seth) to make people aware of climate issues.

Kaime Silvestre Silva Oliveira (T: kaimesilvestres; IG: kaimesilvestre; F: kaime.silvestre) is a Brazilian climate activist and human rights lawyer. Born in the Amazon, he relentlessly advocates for humanitarian action and for protection of the Amazon, its wildlife, and the people who depend on it.

Roman Stratkötter (T: stratkoetter) is an activist who has been striking for SaveCongoRainforest on Twitter since March 19, 2020. He is also an activist for #SaveBugomaForest, AfricaIsNotADumpster, freeNavalny, SaveSahel, StopShell, PeaceUponYemen, WithDrawTheCap, IStandWithTheFarmers, LeilãoFóssilNão/EndFossilFuels, BlackLivesMatter, and loveislove.

Leah Thomas (T: Leahtommi; IG: greengirlleah; F: intersectionalenvironmentalist; W: intersectionalenvironmentalist.com) is an intersectional environmental activist and eco-communicator based in Southern California. She's passionate about advocating for and exploring the relationship between social justice and environmentalism and is the founder of the Intersectional Environmentalist Platform.

Elizabeth Wathuti (T: lizwathuti; IG: lizwathuti; W: lizmazingira.com) is an environmentalist and climate activist from Kenya, and the Founder of Green Generation Initiative. She is currently the Head of Campaigns and the Daima Green Spaces Coalition Coordinator at the Maathai Foundation.

Elizabeth received the Wangari Maathai Scholarship Award from the Green Belt Movement, the Kenya Community Development Foundation, and Rockefeller Foundation in 2016. She has also served as the Chairperson of Kenyatta University Environmental Club and holds a bachelor's degree in Environmental Studies and Community Development from Kenyatta University.

Climate Activists Mentioned in This Book

Isabelle Axelsson (T: isabelle_ax; IG: isabelleax_)

Xiye Bastida (T: xiyebastida; IG: xiyebeara)

Sascha Blidorf (T: SascharBlidorf; IG: saschablidorf)

Connor Childs (W: plasticfreecayman.com/youth-action)

Chibeze Ezekiel (T: chibeze1; IG: chibeze1)

Brianna Fruean (T: Brianna_Fruean; IG: briannafruean)

Hindou Oumarou Ibrahim (T: hindououmar;
  IG: hindououma)

Eva Jones (I: evaastrid37)

Licypriya Kangujam (T: LicypriyaK; IG: licypriyakangujam)

Elizabeth Gulugulu Machache (T: lizgulaz; IG: lizgulaz)

Arshak Makichyan (T: MakichyanA; IG: makichyan.arshak)

Jamie Margolin (T: Jamie_Margolin; IG: jamie_s_margolin)

Steff Mcdermot (W: plasticfreecayman.com/author/
  steffmcdermont/)

Ndoni Mcunu (T: ndonimcunu; IG: ndonimcunu;
  W: ndonimcunu.com)

Ayakha Melithafa (T: AyakhaMelithafa; IG: ayakhamelithafa)

Nyombi Morris (T: mnyomb1; IG: mnyomb1)

Makenna Muigai (T: MakennaMuigai)

Natasha Mwansa (T: TashaWangMwansa; IG: natashamwansa)

Joan & Clare (T: joanandclare1; IG: joan.and.clare)

Luisa Neubauer (T: Luisamneubauer; IG: luisaneubauer)

Sadrach Nirere (T: SadrachNirere; IG: sadrachnirere)

Disha Ravi (T: disharavii; IG: disharavii)

Davis Reuben Sekamwa (T: davisreuben3; IG: davisreuben9.0)

Sasha Shugai (T: sasha_shuga; IG: sashashu__; W: linktr.ee/
sashashugai)

Greta Thunberg (T: gretathunberg; IG: gretathunberg)

Loukina Tille (T: loukinatille; IG: loukinatille)

Amelia "Lia" Tuifua (F: Lia Tuifua)

Alexandria Villaseñor (T: AlexandriaV2005; T: Earth_
Uprising; IG: alexandriav2005)

Brix and Max Whiteman-Muller (F: dirk_dirkman.5)

Remy Zahiga (T: Remy_Zahiga; IG: remyzahiga)

Wenying Zhu (T: Wenying_Z)

Olivia Zimmer (W: plasticfreecayman.com/youth-action)

Books and Documentaries

Hammond, Mel. *Love the Earth: Understanding Climate
Change, Speaking Up for Solutions, and Living an
Earth-Friendly Life* (New York: American Girl, 2020).

Hill, Jordan (dir). *Greta Thunberg: Rebel with a Cause* (2020).

Johnson, Ayana Elizabeth, and Katharine K. Wilkinson (eds).
*All We Can Save: Truth, Courage, and Solutions for the
Climate Crisis* (New York: One World, 2020).

Maathai, Wangari. *The Challenge for Africa* (New York:
Vintage, 2008).

——. *Unbowed: A Memoir* (New York: Vintage, 2006).

Margolin, Jamie. *Youth to Power: Your Voice and How to Use It*
(New York: Hachette, 2020).

Thunberg, Greta. *No One Is Too Small to Make a Difference*
(New York: Penguin, 2018/19).

Slogans for Signs

1.5°C = Rich Countries, Do Your Fair Share

Are You Fracking Kidding Me?

Be Part of the Solution Not Part of the Pollution

Black Lives Matter

Change the System Not the Climate

Climate Change Is Not Fiction

Climate Change Is Worse Than Homework

Educate Girls for Climate

EU, We Are Watching You

Face the Climate Emergency

Hands Off the Arctic

If You Don't Act Like Adults, We Will

I Want You to Panic

Life in Plastic Is Not Fantastic

No Climate Justice Without Racial Justice

People Over Profit

Planet Above Profit

Save the Congo Rainforest

Sea Levels Are Rising and So Are We

Small Acts When Multiplied By Millions of People Are Able to
Transform the World

Stop Destroying Our Planet

There Is No Planet B

There Is No Such Thing As Clean Fossil Fuels

Unite Behind the Science

We Cannot Eat Coal and We Cannot Drink Oil

You Say You Love Your Children, But You're Destroying Their
   Future

Hashtags

#BlackLivesMatter
#ClimateCrisis
#ClimateStrike
#CongoEnviroVoice
#EndPlasticPollution
#FaceTheClimateEmergency
#FightClimateInjustices
#Fightfor1Point5
#FridaysForFuture
#GlobalClimateJustice
#IndigenousLivesMatter
#JustRecovery
#NoMoreEmptyPromises
#RiseUpMovement
#SaveBugomaForest
#SaveCongoRainforest
#SchoolStrike4Climate
#SDGs
#StopEACOP
#WomenRiseUp
#WomensLivesMatter

Notes

Introduction

1 International Energy Agency, *Africa Energy Outlook 2019: World Energy Outlook*, special report (Paris: November 2019), https://www. iea.org/reports/africa-energy-outlook-2019

2 Oxfam: https://oxfamapps.org/media/press_release/average-brit-will-emit-more-by-12-january-than-residents-of-seven-african-countries-do-in-a-year/

3 Associated Press, "Africa Shouldn't Need to Beg for Climate Aid: Bank President," *Arab News*, February 11, 2020, https://www.arabnews.com/node/1626486/business-economy

Chapter 1: Finding My Cause

1 Real Ombuor, "East Africa Flood Deaths Surpass 400," *Voice of America News*, May 24, 2018, https://www.voanews.com/africa/east-africa-flood-deaths-surpass-400; UN News, "Almost 500,000 Affected as Devastating Floods Inundate Central Somalia—UN Mission," May 1, 2018, https://news.un.org/en/story/2018/05/1008612

2 Richard Davis, "Uganda—Dozens Killed in Landslides and Floods in Eastern Region," FloodList, October 12, 2018, http://floodlist.com/africa/uganda-landslides-floods-bududa-eastern-region-october-2018; ReliefWeb, "ACAPS Briefing Note: Uganda—Flooding and Landslides in Bududa District, October 18, 2018, https://reliefweb.int/report/uganda/acaps-briefing-note-uganda-flooding-and-landslides-bududa-district-18-october-2018; Samuel Okiror, "At Least 36 Dead in Uganda

Landslides as School Disappears Beneath Mud," *Guardian*, October 12, 2018, https://www.theguardian.com/global-development/2018/oct/12/uganda-landslides-36-dead-school-disappears-beneath-mud-bududa

3 Derick Msafiri and Ronald Makanga, "AD303: Most Ugandans See Worsening Drought, Say Climate Change Is Making Life Worse," Afrobarometer, Dispatches no. 303, 2019, https://afrobarometer.org/publications/ad303-most-ugandans-see-worsening-drought-say-climate-change-making-life-worse

4 World Meteorological Association, "New Climate Predictions Assess Global Temperatures in Coming Five Years," July 8, 2020, https://public.wmo.int/en/media/press-release/new-climate-predictions-assess-global-temperatures-coming-five-years

5 Daniel Moritz-Rabson, "Temperatures Could Rise up to 7 Degrees Celsius Above Pre-Industrial Levels, Startling Study Shows," *Newsweek*, September 17, 2019, https://www.newsweek.com/climate-change-temperature-rise-seven-degrees-new-un-report-1459666; UN News, "UN Emissions Reports: World on Course for More Than 3 Degree Spike, Even If Climate Commitments Are Met," November 26, 2019, https://news.un.org/en/story/2019/11/1052171

6 "Is This How You Feel," https://www.isthishowyoufeel.com/this-is-how-scientists-feel.html

7 In most families, when a child is born, the first child's last name will come from the paternal grandmother; for the second child, the paternal grandfather will provide the name. In this way, different elders in the family name the latest generation. Many communities in Uganda possess totems that represent their identity. When people in my community, which is the Baganda, hear my last name, Nakate, they associate me with our totem, which is a cow, or *ente* or *akate* in Luganda. The Baganda, like other such communities, are divided into clans. Mine is the Njovu, or the elephant. My two sisters and two brothers have last names from that clan, and all naming comes entirely from the paternal side.

Chapter 2: Striking Out

1 Jonah Kirabo, "Climate Activists Arrested for Holding Demo Outside Parliament," *Nile Post*, February 27, 2021, https://nilepost.co.ug/2021/02/27/climate-activists-arrested-for-holding-demo-outside-parliament/

Chapter 3: COP Out

1 FloodList, "Uganda—8 Dead as Flash Floods Hit Kampala," May 30, 2019, http://floodlist.com/africa/uganda-flash-floods-kampala-may-2019

2 FloodList, "Uganda—Deadly Floods and Landslides in Eastern Region (Updated)," December 5, 2019, http://floodlist.com/africa/uganda-floods-bududa-sironko-december-2019

3 FloodList, "Uganda—Hundreds Homeless After Floods in Eastern and Western Regions," October 21, 2019, http://floodlist.com/africa/uganda-floods-eastern-western-region-october-2019

4 FloodList, "Uganda—6 Killed in Floods and Landslides After More Heavy Rain," November 6, 2019, http://floodlist.com/africa/uganda-floods-landslides-november-2019

Chapter 4: Crop Out

1 National Snow & Ice Data Center, "Climate Change in the Arctic," (updated May 4, 2020), https://nsidc.org/cryosphere/arctic-meteorology/climate_change.html

2 British Antarctic Survey, "Past Evidence Supports Complete Loss of Arctic Sea-ice by 2035," *Science Daily*, August 10, 2010, https://www.sciencedaily.com/releases/2020/08/200810113216.htm

3 Lauren Easton, "AP Statement on Cropped Photo," AP.org, January 24, 2020, https://blog.ap.org/announcements/ap-statement-on-cropped-photo

4 Kenya Evelyn, "Outrage at Whites-Only Image as Ugandan Climate Activist Cropped from Photo," *Guardian*, January 25, 2020, https://www.theguardian.com/world/2020/jan/24/whites-only-photo-uganda-climate-activist-vanessa-nakate

5 Jamey Keaten and Pan Pylas, "Thunberg Brushes Off Mockery from US Finance Chief," AP, January 24, 2020, https://apnews.com/article/ee36c1b18874d3ebec2c743f0968396f. The picture was replaced with the photo from the press conference.

6 BBC, "Vanessa Nakate: Climate Activist Hits Out at 'Racist' Photo Crop," January 24, 2020, https://www.bbc.com/news/world-africa-51242972

7 YouTube, "COP25 Speech | Hilda Flavia Nakabuye," April 9, 2020, https://www.youtube.com/watch?v=wgpYF9iVotg

8 Quoted in Kenya Evelyn, " 'Like I Wasn't There': Climate Activist Vanessa Nakate on Being Erased from a Movement," *Guardian*, January

29, 2020, https://www.theguardian.com/world/2020/jan/29/
vanessa-nakate-interview-climate-activism-cropped-photo-davos

9 Chelsea McFadden, "Vanessa Nakate and Perceptions of Black Student
Activists," *Journal of Sustainability Education*, December 29, 2020,
http://www.susted.com/wordpress/content/vanessa-nakate-and-
perceptions-of-black-student-activists_2020_12/

10 Leigh Haber, "Ugandan Climate Activist Vanessa Nakate Will Release
Her First Book This Fall," *Oprah*, January 25, 2021, https://www.
oprahdaily.com/entertainment/books/amp35293052/vanessa-nakate-a-
bigger-picture-book-interview/

11 YouTube, "Press & Racism with Vanessa Nakate," July 2, 2020, https://
www.youtube.com/watch?v=9juD2HImo2Q&t=237s

12 Kenya Evelyn, "'Like I Wasn't There.'"

13 David Bauder, "Photo Cropping Mistake Leads to AP Soul-Searching
on Race," AP, January 27, 2020, https://apnews.com/article/6a853a81
f34164ab85713e68a889976d

14 Al Jazeera, "Anger as Ugandan Activist Cropped Out of Photo with
White Peers," January 25, 2020, https://www.aljazeera.com/news/
2020/1/25/anger-as-ugandan-activist-cropped-out-of-photo-with-
white-peers.

15 Amy Woodyatt, "South Africans Outraged as US Journalist Describes
President as 'Unidentified Leader,'" CNN, August 27, 2019, https://
www.cnn.com/2019/08/27/africa/south-africa-unidentified-leader-intl-
scli/index.html

Chapter 5: We Are All Africa

1 World Economic Forum, "Are Forests Carbon Sinks or Carbon
Sources," *EcoWatch*, February 15, 2020, https://www.ecowatch.com/
forests-climate-change-2650542772.html

2 Morgan Erickson-Davis, "Congo Basin Rainforest May Be Gone by
2100, Study Finds," *Mongabay*, November 7, 2018, https://news.
mongabay.com/2018/11/congo-basin-rainforest-may-be-gone-by-
2100-study-finds

3 Rhett A. Butler, "How the Pandemic Impacted Rainforests in 2020:
A Year in Review," *Mongabay*, December 28, 2020, https://news.
mongabay.com/2020/12/how-the-pandemic-impacted-rainforests-in-
2020

4 Rhett A. Butler, "Global Forest Loss Increases in 2020," *Mongabay*,
March 31, 2021, https://news.mongabay.com/2021/03/global-forest-
loss-increases-in-2020-but-pandemics-impact-unclear

5   Greenpeace Africa, "#SaveCongoRainforest Zoom Action on World
    Biodiversity Day," May 22, 2020, https://www.greenpeace.org/africa/en/
    publications/11202/savecongorainforest-zoom-action

6   Charlotte Edmond, "Cape Town Almost Ran Out of Water. Here's How
    It Averted the Crisis," World Economic Forum, August 23, 2019,
    https://www.weforum.org/agenda/2019/08/cape-town-was-90-days-
    away-from-running-out-of-water-heres-how-it-averted-the-crisis

7   ReliefWeb, "2018–2019 Mozambique Humanitarian Response Plan
    Revised Following Cyclones Ida and Kenneth, May 2019 (November
    2018–June 2019)," May 25, 2019, https://reliefweb.int/report/
    mozambique/2018-2019-mozambique-humanitarian-response-plan-
    revised-following-cyclones-idai

8   ReliefWeb, "Niger: Situation Report, 26 September 2019," September
    26, 2019, https://reliefweb.int/report/niger/
    niger-situation-report-26-sep-2019.

9   FloodList, "Niger—Floods Leave 43 Dead and 5,000 Homes Destroyed,"
    September 5, 2019, http://floodlist.com/africa/niger-floods-september-
    2019

10 Laureen Fagan, "International Aid Arrives as Djibouti Cleans Up from
    Floods," *Africa Times*, November 30, 2019, https://africatimes.com/
    2019/11/30/international-aid-arrives-as-djibouti-cleans-up-from-
    floods

11 AT Editor, "Kenya Landslide Toll at 37 as Rains Continue Over Africa,"
    Africa Times, November 23, 2019, https://africatimes.com/2019/11/23/
    kenya-landslide-toll-at-37-as-rains-continue-over-east-africa

12 Reuters, "Ugandan Hospital, Somali Town Washed Away by East Africa
    Floods," May 8, 2020, https://news.trust.org/item/20200508125400-
    wf2b1/; https://www.climate-refugees.org/spotlight/5/11/2020-1

13 Madeline Stone, "A Plague of Locusts Has Descended on East Africa.
    Climate Change May Be to Blame," *National Geographic*, February 14,
    2020, https://www.nationalgeographic.com/science/article/locust-
    plague-climate-science-east-africa

14 Samuel Egadu Okiror, "Uganda Faces Food Shortage as Coronavirus
    Disrupts Locust Fight," *Al Jazeera*, April 9, 2020, https://www.aljazeera.
    com/news/2020/4/9/uganda-faces-food-shortage-as-coronavirus-
    disrupts-locust-fight

15 Dylan Barth and Mark Abadi, "Swarms of Locusts Have Destroyed
    170,000 Acres of Crops in East Africa—and Local Farmers Are
    Nearly Helpless to Stop It," *Business Insider*, February 25, 2020,
    https://www.businessinsider.com/locusts-africa-kenya-farmers-crops-
    2020-2

16 Evan Girvetz, Julian Ramirez-Villegas, Lieven Claessens, et al. "Future Climate Projections in Africa: Where Are We Headed?" *The Climate-Smart Agriculture Papers*, November 28, 2018, https://link.springer.com/chapter/10.1007/978-3-319-92798-5_2.

17 Future Climate for Africa, *Future Climate for Africa*, 2nd Edition 2017, https://futureclimateafrica.org/wp-content/uploads/2016/01/cdkj5678_fcfa_brochure_2ndedition_1708_web.pdf.

18 I. Niang, O. C. Ruppel, M.A. Abdrabo, et al. 2014: Africa. In: *Climate Change 2014: Impacts, Adaptation, and Vulnerability. Part B: Regional Aspects*. Contribution of Working Group II to the Fifth Assessment Report of the Intergovernmental Panel on Climate Change [Barros, V.R., C.B. Field, D.J. Dokken, M.D. Mastrandrea, K.J. Mach, T.E. Bilir, M. Chatterjee, K.L. Ebi, Y.O. Estrada, R.C. Genova, B. Girma, E.S. Kissel, A.N. Levy, S. MacCracken, P.R. Mastrandrea, and L.L.White (eds.)]. Cambridge University Press, Cambridge, United Kingdom, and New York, NY, USA, pp. 1199–1265.

19 Girvetz, Ramirez-Villegas, Claessens, et al. "Future Climate Projections."

20 World Wildlife Fund, "Backgrounder: Regional Impacts + the 1.5°C Climate Target—Africa," https://wwfeu.awsassets.panda.org/downloads/backgrounder___africa_at_1_5c.pdf (accessed April 19, 2021).

21 Tom K. R. Matthews, Robert L. Wilby, and Conor Murphy, "Communicating the Deadly Consequences of Global Warming for Human Heat Stress," *PNAS*, March 27, 2017, https://doi.org/10.1073/pnas.1617526114

22 Raluca Besliu, "Togo's Battle with Coastal Erosion," DW, April 17, 2017, https://www.dw.com/en/togos-battle-with-coastal-erosion/a-38378211.

23 Binh Pham-Duc, Florence Silvestre, Fabrice Papa, et al. "The Lake Chad Hydrology Under Current Climate Change," *Scientific Reports* 10: 5498 (2020), https://doi.org/10.1038/s41598-020-62417-w

24 Will Ross, "Lake Chad: Can the Vanishing Lake Be Saved?" BBC News, March 31, 2018, https://www.bbc.com/news/world-africa-43500314

Chapter 6: A Greener Uganda

1 YouTube, "C40 World Mayors Summit 2019: Hilda Flavia Nakabuye, October 11, 2019, https://www.youtube.com/watch?v=OF7vT3cmC3g

2 M. Josephat, "Deforestation in Uganda: Population Increase, Forests Loss and Climate Change," *Environmental Risk Assessment and Remediation* 2018 2(2):46–50, DOI: 10.4066/2529-8046.100040.

3   Nkulumo Zinyengere, Julio Araujo, John H. Marsham, and David
    P. Rowell, "Uganda Country Fact Sheet: Current and Projected Future
    Climate," in *Africa's Climate: Helping Decision-Makers Make Sense of
    Climate Information*, Leonie Joubert (ed). (London: Climate &
    Development Knowledge Network, 2016), pp. 92–100, https://www.
    researchgate.net/publication/311455953_uganda_country_fact_sheet_
    current_and_projected_future_climate

4   M. Josephat, "Deforestation in Uganda."

5   Victor Raballa, "Experts Call for Action Over Rising Lake Victoria
    Water Levels," *The East African*, May 18, 2020, https://www.
    theeastafrican.co.ke/news/ea/Lake-Victoria-water-rises-to-historic-
    levels/4552908-5556404-n0j7crz/index.html

6   Hope Mafaranga, "Heavy Rains, Human Activity, and Rising Waters at
    Lake Victoria," *Eos*, July 7, 2020, https://eos.org/articles/
    heavy-rains-human-activity-and-rising-waters-at-lake-victoria

7   National Forestry Authority, "Surrender Illegal Land Titles in Forest
    Reserves," February 26, 2019, https://www.nfa.go.ug/index.php/
    12-nfa-news

8   Liam Taylor, " 'Cutting Everything in Sight': Ugandans Vow to Curb
    Rampant Deforestation," March 11, 2019, https://www.reuters.com/
    article/us-uganda-landrights-deforestation/cutting-everything-in-
    sight-ugandans-vow-to-curb-rampant-deforestation-IDUSKBN1
    QT03E

9   Global Forest Watch, "Primary Forest Loss in Uganda," https://www.
    globalforestwatch.org/dashboards/country/UGA (accessed April 19,
    2021).

10 Global Environment Facility, "Greening Charcoal Production in
    Uganda," October 7, 2018, https://www.thegef.org/news/greening-
    charcoal-production-uganda#

11 Obinna Ekeh, Andreas Fangmeier, and Joachim Müller, "Quantifying
    Greenhouse Gases from the Production, Transportation and
    Utilization of Charcoal in Developing Countries: A Case Study of
    Kampala, Uganda," *The International Journal of Life Cycle Assessment* 19:
    1643–1652 (2014), https://link.springer.com/article/10.1007/s11367-
    014-0765-7

12 M. Josephat, "Deforestation in Uganda."

13 The World Bank, "Uganda Economic Update Recommends Expanding
    Social Protection Programs to Boost Inclusive Growth," February 13,
    2020, https://www.worldbank.org/en/news/press-release/2020/02/13/
    uganda-economic-update-recommends-expanding-social-protection-
    programs-to-boost-inclusive-growth

14 WFP, "WFP Uganda Country Brief," November 2020, https://docs.wfp.
    org/api/documents/WFP-0000123695/download

15 FAO, *Africa Sustainable Livestock 2050: The Future of Livestock in
    Uganda: Opportunities and Challenges in the Face of Uncertainty* (Rome:
    Food and Agriculture Organization of the United Nations, 2019),
    http://www.fao.org/3/ca5420en/CA5420EN.pdf

16 Dastan Bamwesigye, Petr Kupec, and Georges Chekuimo, et al.,
    "Charcoal and Wood Biomass Utilization in Uganda: The
    Socioeconomic and Environmental Dynamics and Implications,"
    Sustainability 2020, 12, 8337; doi:10.3390/su12208337.

17 Clean Cooking Alliance, "Uganda," http://cookstoves-admin.digitopia.
    net/country-profiles/focus-countries/8-uganda.html (accessed
    September 14, 2021).

18 First Climate, "Uganda: Avoided Deforestation Using Efficient
    Cookstoves," https://www.firstclimate.com/en/our-carbon-offset-
    project/uganda-avoided-deforestation-using-efficient-cookstoves
    (accessed April 19, 2021).

19 Shuaib Lwasa, "Uganda Offers Lessons in Tapping the Power of Solid
    Waste," Phys.Org, September 3, 2019, https://phys.org/news/2019-09-
    uganda-lessons-power-ofsolidwaste.html

20 IndustriALL Global Union, "Ugandan Oil and Gas Fields Provide
    Potential for Union Organizing," October 1, 2020, http://www.
    industriall-union.org/ugandan-oil-and-gas-fields-provide-potential-
    for-union-organizing

21 Oil & Gas Journal, "Uganda Approves FEED, EPC Contractor for
    Proposed Refinery," March 13, 2019, https://www.ogj.com/refining-
    processing/article/17278792/uganda-approves-feed-epc-contractor-
    for-proposed-refinery

22 Charné Hundermark, "Ugandan Government Issues New Timelines for
    Oil Refinery Project," Africa Oil & Power, August 19, 2020, https://
    www.africaoilandpower.com/2020/08/19/ugandan-government-issues-
    new-timelines-for-oil-refinery-project

23 IndustriALL Global Union, "Ugandan Oil and Gas Fields."

24 Fred Pearce, "A Major Oil Pipeline Project Strikes Deep at the Heart of
    Africa," YaleEnvironment360, May 21, 2020, https://e360.yale.edu/
    features/a-major-oil-pipeline-project-strikes-deep-at-the-heart-of-
    africa

25 Charity Migwi, Edwin Mumbere, and Evelyn Acham, "East African
    Crude Oil Pipeline (EACOP) Will Disenfranchise Local Communities
    in Uganda and Tanzania," *Africa News*, March 26, 2021, https://www.
    africanews.com/2021/03/26/east-african-crude-oil-pipeline-eacop-

will-disenfranchise-local-communities-in-uganda-and-tanzania-by-charity-migwi-edwin-mumbere-and-evelyn-acham

26 The World Bank, "Access to Electricity, Rural (% of Rural Population)—Uganda," https://data.worldbank.org/indicator/EG.ELC.ACCS.RU.ZS?locations=UG (accessed April 19, 2021).

27 Sari Fordham, "Adventist Environmentalist Tackles Plastic Pollution in Uganda," *The Spectrum*, February 10, 2021, https://spectrummagazine.org/interviews/2021/adventist-environmentalist-tackles-plastic-pollution-uganda

28 End Plastic Pollution Now, https://endplasticpollutionnow.blogspot.com (accessed April 19, 2021).

29 UN News, "Food Systems Account for Over One-Third of Global Greenhouse Gas Emissions," March 9, 2021, https://news.un.org/en/story/2021/03/1086822

30 FAO, "Key Facts and Findings," http://www.fao.org/news/story/en/item/197623/icode/ (accessed April 19, 2021).

Chapter 7: Speaking Out for Women and Girls

1 CAMFED, "Education Is a Universal Right. It Is Also a Matter of Justice," https://camfed.org/why-girls-education/ (accessed April 19, 2021).

2 UNICEF, "Girls' Education," https://www.unicef.org/education/girls-education (accessed April 19, 2021).

3 Project Drawdown, "Table of Solutions," https://drawdown.org/solutions/table-of-solutions (accessed April 19, 2021). This analysis models the impact of increased adoption of family planning from 2020 to 2050 on emissions from energy use, building space, food, waste, and transportation by comparing two scenarios: a high adoption scenario, in which there is increased adoption of family planning, and a reference scenario with no additional investment in family planning. Educational attainment also influences the dynamics of fertility and therefore projections of future population growth. Among educational factors influencing population growth, universal access to and equal quality of education, as well as reproductive health-specific education are predominant. In most contexts, education influences timing of marriage, timing of births, desired family size, and total number of births.

4 Vanessa Nakate, "Educating Young Women is the Climate Fix No One Is Talking About," *Wired UK*, January 27, 2021, https://www.wired.co.uk/article/educating-girls-climate-change

5 UNICEF, "Child Marriage," April 2020, https://data.unicef.org/topic/child-protection/child-marriage

6 Uganda For Her, "Girls' Education," https://uganda4her.org/girls-education (accessed April 19, 2021).

7 UNICEF, "Child Marriage."

8 The World Bank, "Maternal Mortality Ratio (Modeled Estimate, per 100,000 Live Births," https://data.worldbank.org/indicator/SH.STA.MMRT (accessed April 19, 2021).

9 UNICEF, "Child Marriage."

10 The World Bank, "Educating Girls and Ending Child Marriage: A Priority for Africa (English)," Working Paper, November 19, 2018, https://documents.worldbank.org/en/publication/documents-reports/documentdetail/268251542653259451/educating-girls-and-ending-child-marriage-a-priority-for-africa

11 UNFPA, "Youth Voices: Securing the Future of Women in Africa by Standing with Girls Today," December 19, 2018, https://reliefweb.int/report/world/youth-voices-securing-future-women-africa-standing-girls-today

12 Amnesty International, "Troll Patrol Findings," https://decoders.amnesty.org/projects/troll-patrol/findings (accessed April 19, 2021).

13 UNFCCC, "Climate Change Increases the Risk of Violence Against Women," November 25, 2019, https://unfccc.int/news/climate-change-increases-the-risk-of-violence-against-women

14 Monica Campo and Sarah Tayton, "Domestic and Family Violence in Regional, Rural and Remote Communities: An Overview of Key Issues," Child Family Community Australia, December, 2015. https://aifs.gov.au/cfca/sites/default/files/publication-documents/cfca-resource-dv-regional.pdf.

15 UN Women, "The Shadow Pandemic: Violence Against Women During COVID-19," 2020. https://www.unwomen.org/en/news/in-focus/in-focus-gender-equality-in-COVID-19-response/violence-against-women-during-COVID-19

16 UN Women, "Facts and Figures: Ending Violence Against Women," https://www.unwomen.org/en/what-we-do/ending-violence-against-women/facts-and-figures (accessed April 19, 2021).

Chapter 8: Rise Up for Justice

1 BBC, "Racism Against Black People in EU 'Widespread and Entrenched,'" November 28, 2018, https://www.bbc.com/news/world-europe-46369046

2 Nan DasGupta, Vinay Shandal, Daniel Shadd, et al. "The Pervasive Reality of Anti-Black Racism in Canada," Boston Consulting Group, December 14, 2020, https://www.bcg.com/en-ca/publications/2020/reality-of-anti-black-racism-in-canada

3 Raphael Tsavkko Garcia, "Diversity in Brazil Is Still Just an Illusion," *Al Jazeera*, October 22, 2020, https://www.aljazeera.com/opinions/2020/10/22/diversity-in-brazil-is-still-just-an-illusion

4 Andreas Illmer, "Black Lives Matter Pushes Japan to Confront Racism," *BBC News*, August 28, 2020, https://www.bbc.com/news/world-asia-53428863

5 Abdi Latif Dahir, Ruth Maclean, and Lynsey Chutel, "George Floyd's Killing Prompts Africans to Call for Police Reform at Home," *New York Times*, July 3, 2020, https://www.nytimes.com/2020/07/03/world/africa/george-floyd-protests-police-africa.html

6 Libby George, "Black Lives Matter Co-Founder Urges Nigeria to Free Jailed Police Protesters," Reuters, December 10, 2020, https://www.reuters.com/article/us-nigeria-protests-blm-idUSKBN28K0UQ.

7 Eoin McSweeney, "Alicia Keys, Greta Thunberg and Others Urge Nigeria to Free Protesters," *CNN*, December 10, 2020, https://www.cnn.com/2020/12/10/africa/celebrities-buhari-letter-endsars-intl/index.html

8 Stanford Medicine, "More Than Half of In-Hospital Deaths from COVID-19 Among Black, Hispanic Patients, Study Finds," November 17, 2020, https://med.stanford.edu/news/all-news/2020/11/deaths-from-covid-19-of-inpatients-by-race-and-ethnicity.html; Tony Kirby, "Evidence Mounts on the Disproportionate Effect of COVID-19 on Ethnic Minorities," *The Lancet*, May 8, 2020, https://www.thelancet.com/journals/lanres/article/PIIS2213-2600(20)30228-9/fulltext; Associated Press, "French Coronavirus Study Finds Black Immigrant Deaths Doubled at Peak," *Guardian*, July 7, 2020, https://www.theguardian.com/world/2020/jul/07/french-coronavirus-study-finds-black-immigrant-deaths-doubled-at-peak; Victoria Waldersee, "COVID Toll Turns Spotlight on Europe's Taboo of Data by Race," Reuters, November 19, 2020, https://www.reuters.com/article/uk-health-coronavirus-europe-data-insigh-idUKKBN27Z0K6; The Conversation, "COVID-19 Is Deadlier for Black Brazilians, A Legacy of Structural Racism that Dates Back to Slavery," June 10, 2020, https://theconversation.com/COVID-19-is-deadlier-for-black-brazilians-a-legacy-of-structural-racism-that-dates-back-to-slavery-139430

9 Miriam Fauzia, "Fact Check: Coronavirus Deaths Across Continent Are Far Less than in U.S.," *USA Today*, November 30, 2020, https://

www.usatoday.com/story/news/factcheck/2020/11/30/fact-check-coronavirus-deaths-africa-far-less-than-u-s/3766222001

10   BBC News, "Ella Adoo-Kissi-Debra: Air Pollution a Factor in Girl's Death, Inquest Finds," December 16, 2020, https://www.bbc.com/news/uk-england-london-55330945

11   Stefano Valentino, "London the Worst City in Europe for Health Costs from Air Pollution," *Guardian*, October 21, 2020, https://www.theguardian.com/environment/2020/oct/21/london-the-worst-city-in-europe-for-health-costs-from-air-pollution

12   CE Delft, "Health Costs of Air Pollution in European Cities and the Linkage with Transport (Report from the European Public Health Alliance," European Commission, October 27, 2020, https://ec.europa.eu/jrc/communities/en/community/city-science-initiative/document/health-costs-air-pollution-european-cities-and-linkage

13   Luiz Sanchez, "Air Pollution Costs Egypt 3.58% of GDP in Welfare Losses," *Madamasr*, September 8, 2016, https://www.madamasr.com/en/2016/09/08/news/u/air-pollution-costs-egypt-3-58-of-gdp-in-welfare-losses

14   Panle Jia Barwick, Shanjun Li, Deyu Rao, Nahim Bin Zahur, "The Impact of Air Pollution on Healthcare Spending in China," *VoxDev*, April 18, 2019, https://voxdev.org/topic/health-education/impact-air-pollution-healthcare-spending-china

15   Elisheva Mittelman, "Air Pollution from Fossil Fuels Costs $8 Billion Per Day, new Research Finds," YaleEnvironment360, https://e360.yale.edu/digest/air-pollution-from-fossil-fuels-costs-8-billion-per-day-new-research-finds

16   NDTV, "Over 50,000 People in Delhi Died Due to Air Pollution Last Year: Study," February 18, 2001, https://www.ndtv.com/delhi-news/over-50-000-people-in-delhi-died-due-to-pm2-5-air-pollution-last-year-study-2373223

17   Human Rights Watch, " 'The Air Is Unbearable,' " August 26, 2020, https://www.hrw.org/report/2020/08/26/air-unbearable/health-impacts-deforestation-related-fires-brazilian-amazon

18   Nathan Rott, "Study Finds Wildfire Smoke More Harmful to Humans Than Pollution from Cars," *NPR*, March 5, 2021, https://www.npr.org/sections/health-shots/2021/03/05/973848360/study-finds-wildfire-smoke-more-harmful-to-humans-than-pollution-from-cars

19   Jessica Learish, "The Most Polluted Cities in the World, Ranked," *CBS News*, August 23, 2019, https://www.cbsnews.com/pictures/the-most-polluted-cities-in-the-world-ranked/36

20 Leah Burrows, "Deaths from Fossil Fuel Emissions Higher Than
    Previously Thought," Harvard School of Engineering, February 9, 2021,
    https://www.seas.harvard.edu/news/2021/02/deaths-fossil-fuel-
    emissions-higher-previously-thought

21 James Hitchings-Hales, "Fossil Fuels Responsible for 1 in 5 of All
    Global Deaths in 2018: Report," Global Citizen, February 9, 2021,
    https://www.globalcitizen.org/en/content/fossil-fuels-air-pollution-1-
    in-5-global-deaths

22 UN News, "Environmental Racism in Louisiana's 'Cancer Alley' Must
    End, Say UN Human Rights Experts," March 2, 2021, https://news.
    un.org/en/story/2021/03/1086172

23 Food Empowerment Project, "Environmental Racism," https://
    foodispower.org/environmental-and-global/environmental-racism/
    (accessed April 19, 2021).

24 NRDC, "Flint Water Crisis: Everything You Need to Know," November
    8, 2018, https://www.nrdc.org/stories/flint-water-crisis-everything-
    you-need-know

25 National Farm Worker Ministry, "Health & Safety," http://nfwm.org/
    farm-workers/farm-worker-issues/health-safety (accessed April 19,
    2021).

26 Nathalie Baptiste, "Farmworkers Are Dying from Extreme Heat,"
    Mother Jones, August 24, 2018, https://www.motherjones.com/
    food/2018/08/farmworkers-are-dying-from-extreme-heat

27 Get Invest, "Zambia: Energy Sector," https://www.get-invest.eu/market-
    information/zambia/energy-sector (accessed April 19, 2021).

28 Burak Bir, "Environmental Disasters Across World in February," *Anadolu
    Agency*, March 1, 2020, https://www.aa.com.tr/en/environment/
    environmental-disasters-across-world-in-february/1750530

Chapter 9: Forecast: Emergency

1 More details and information on the UN Sustainable Development
   Goals can be found at: https://sdgs.un.org/goals and https://unstats.
   un.org/sdgs

2 Our World in Data, "Global Number Affected by Natural Disasters, All
   Natural Disasters, 1900 to 2019," https://ourworldindata.org/grapher/
   total-affected-by-natural-disasters (accessed April 19, 2021).

3 Stéphane Hallegatte and Brian Walsh, "COVID, Climate Change and
   Poverty: Avoiding the Worst Impacts," World Bank, October 7, 2020,
   https://blogs.worldbank.org/climatechange/covid-climate-change-and-
   poverty-avoiding-worst-impacts

4   UNICEF, "COVID-19: A Threat to Progress Against Child Marriage," March 2021, https://data.unicef.org/resources/COVID-19-a-threat-to-progress-against-child-marriage

5   United Nations Department of Economic and Social Affairs, "Goal 6," https://sdgs.un.org/goals/goal6

6   World Health Organization, "World Malaria Report 2020," https://www.who.int/teams/global-malaria-programme/reports/world-malaria-report-2020

7   United Nations Department of Economic and Social Affairs, "Goal 15," https://sdgs.un.org/goals/goal15

8   "Climate Justice," Sustainable Development Goals, May 2019, https://www.un.org/sustainabledevelopment/blog/2019/05/climate-justice

9   David Vetter, "Africa Could Be Locked into Fossil Fuel Future, Warns New Report," *Forbes*, January 11, 2021, https://www.forbes.com/sites/davidrvetter/2021/01/11/africa-could-be-locked-into-fossil-fuel-future-warns-new-report

10 Gary Fuller, "Pollutionwatch: Africa Increases Its Reliance on Fossil Fuels," *Guardian*, November 7, 2019, https://www.theguardian.com/environment/2019/nov/07/pollutionwatch-africa-increases-reliance-fossil-fuels#

11 Mining in Africa, "Coal Mining in Africa," https://www.miningafrica.net/natural-resources-africa/coal-mining-in-africa (accessed April 19, 2021).

12 The World Bank, "Climate Finance Drives Action on the Ground," June 15, 2020, https://www.worldbank.org/en/news/feature/2020/06/15/climate-finance-drives-action-on-the-ground

13 Emma Foehringer Merchant, "Report: Banks Have Invested $1.9 Trillion in Fossil Fuels Since 2015," *Green Tech Media*, March 20, 2019, https://www.greentechmedia.com/articles/read/report-banks-have-1-9-trillion-into-fossil-fuel-extraction-since-2015

14 International Renewable Energy Agency, "Global Energy Transformation: A Roadmap to 2050 (2019 Edition), United Arab Emirates, April 2019, https://www.irena.org/publications/2019/Apr/Global-energy-transformation-A-roadmap-to-2050-2019Edition

15 Rainforest Action Network, BankTrack, Indigenous Environmental Network, et al., Banking on Climate Change: Fossil Fuel Finance Report 2020, March 18, 2020, https://www.ienearth.org/banking-on-climate-change-fossil-fuel-report-cards

16 Simon Evans, "The IEA Weighs in on Stranded Assets—Not Just a Green Conspiracy?" CarbonBrief, June 4, 2014, https://www.

carbonbrief.org/the-iea-weighs-in-on-stranded-assets-not-just-a-green-conspiracy#

17 Council on State Fragility, "Powering Up Energy Investments in Fragile States: A Call to Action," International Growth Centre, February 2021, https://www.fragilitycouncil.org/publication/powering-energy-investments-fragile-states-call-action

18 I. Niang, O. C. Ruppel, M.A. Abdrabo, et al. 2014: Africa. In: *Climate Change 2014: Impacts, Adaptation, and Vulnerability. Part B: Regional Aspects.* Contribution of Working Group II to the Fifth Assessment Report of the Intergovernmental Panel on Climate Change [Barros, V.R., C.B. Field, D.J. Dokken, M.D. Mastrandrea, K.J. Mach, T.E. Bilir, M. Chatterjee, K.L. Ebi, Y.O. Estrada, R.C. Genova, B. Girma, E.S. Kissel, A.N. Levy, S. MacCracken, P.R. Mastrandrea, and L.L.White (eds.)]. Cambridge University Press, Cambridge, United Kingdom, and New York, NY, USA, pp. 1199–1265, https://www.ipcc.ch/site/assets/uploads/2018/02/WGIIAR5-Chap22_FINAL.pdf

19 Wangari Maathai, "Protect Human Rights, Protect Planet Rights," A Statement by Wangari Maathai at the Launch of The United Nations Human Rights Council, June 19, 2006, https://www.ohchr.org/Documents/HRBodies/HRCouncil/RegularSession/Session1/wangari_maathai.pdf